Texts and Monographs in
Symbolic Computation

A Series of the
Research Institute for Symbolic Computation,
Johannes-Kepler-University, Linz, Austria

Edited by
B. Buchberger and G. E. Collins

A. Miola and
M. Temperini (eds.)

Advances in the Design of Symbolic Computation Systems

SpringerWienNewYork

Dr. Alfonso Miola
Dipartimento di Informatica e Automazione
Università degli Studi di Roma Tre, Rome, Italy

Dr. Marco Temperini
Dipartimento di Informatica e Sistemistica
Università degli Studi di Roma "La Sapienza", Rome, Italy

Printing was supported by the Fonds zur Förderung der wissenschaftlichen Forschung and by the Linzer Hochschulfonds.

Printed in Austria

Data conversion by H.-D. Ecker, Büro für Textverarbeitung, Bonn
Printed by Novographic, Ing. Wolfgang Schmid, A-1230 Wien
Graphic design: Ecke Bonk
Printed on acid-free and chlorine-free bleached paper

With 39 Figures

Library of Congress Cataloging-in-Publication Data

Advances in the design of symbolic computation systems / A. Miola and M. Temperini (eds.).
p. cm. — (Texts and monographs in symbolic computation, ISSN 0943-853X)
Includes bibliographical references and index.
ISBN 3-211-82844-3 (alk. paper).
1. System design. 2. Object-oriented programming.
3. Mathematics—Data processing. I. Miola, A. (Alfonso), 1944–
II. Temperini, M. (Marco) III. Series.
QA76.9.S88A38 1997
004.2'1—DC21

97-8929
CIP

ISSN 0943-853X
ISBN 3-211-82844-3 Springer-Verlag Wien New York

Preface

This book presents a collection of coordinated scientific papers describing the research activities and the scientific and technical results achieved within the TASSO research initiative. TASSO has been promoted and funded by the Italian National Research Council (Consiglio Nazionale delle Ricerche), within the national research project "Sistemi Informatici e Calcolo Parallelo." Part of the project has been conducted in cooperation with the Research Institute for Symbolic Computation (RISC) of the Johannes-Kepler-University, Linz, Austria. New methodological aspects related to design and implementation of symbolic computation systems are considered in TASSO. The aim of this research project is to integrate such aspects into a homogeneous software environment for scientific computation. The proposed methodology is based on a combination of different techniques: algebraic specification through a modular approach and completion algorithms, approximated and exact algebraic computing methods, object-oriented programming paradigm, automated theorem proving through methods à la Hilbert and methods of natural deduction. In order to experiment with the proposed methodology and to show its effectiveness, the design of a prototypal implementation of a computing environment represents a specific goal of the project. In this environment, several modules for specification of mathematical structures and treatment of their logic properties are integrated and managed by a suitable interface.

While the focus is on the theoretical activity developed in the project, some papers present also practical achievements: the TASSO programming language, implementations of deductive mechanisms, and algebraic computing tools. Some papers not directly connected to the practical achievements have been included, to provide a better comprehension of general arguments related to the topics of the project.

The contents are arranged in four parts, each one collecting and sequencing related subjects.

The first describes the state of the art in symbolic mathematical computation and the requirements for advanced software systems; it motivates and founds the general approach followed by TASSO for the integration of specification and computing paradigms.

The proposed treatment of mathematical objects, via techniques for method abstraction, structures classification, and exact representation is presented in the second part.

The programming methodology that supports the design and implementation issues is described in the third part.

The reasoning capabilities supported by the environment are presented in the fourth part.

Academic partners active in TASSO have been:
Università di Roma "La Sapienza": Dipartimento di Informatica e Sistemistica, Roma, Italy
Università di Roma Tre: Dipartimento di Informatica e Automazione
Università di L'Aquila: Dipartimento di Matematica Pura ed Applicata, L'Aquila, Italy
Consiglio Nazionale delle Ricerche (CNR): Istituto di Analisi dei Sistemi ed Informatica, Roma, Italy
Consiglio Nazionale delle Ricerche (CNR): Istituto per la Matematica Applicata, Genova, Italy
Johannes-Kepler-University: Research Institute for Symbolic Computation (RISC), Linz, Austria
University of Warszaw: Department of Informatics, Warszaw, Poland

Cooperating industrial partners have been:
Olivetti Systems and Networks, Ivrea, Italy
STET: Scuola Speciale G. Reiss Romoli, L'Aquila, Italy

The work for the TASSO project, and the interest in related subjects arisen within the international scientific community, gave great impulse to the organization of the DISCO conferences on "Design and Implementation of Symbolic Computation Systems," held in Capri (Italy, 1990), Bath (United Kingdom, 1992), Gmunden (Austria, 1993), and Karlsruhe (Federal Republic of Germany).

A. Miola and M. Temperini

Contents

List of contributors

S. Antoy, Department of Computer Science, Portland State University, P.O. Box 751, Portland, OR 97207, U.S.A.

P. Bertoli, Dipartimento di Informatica e Automazione, Università degli Studi di Roma Tre, Via Vasca Navale 84, I-00146 Roma, Italy.

B. Buchberger, Research Institute for Symbolic Computation, Johannes-Kepler-University Linz, A-4040 Linz, Austria.

O. Caprotti, Research Institute for Symbolic Computation, Johannes-Kepler-University Linz, A-4040 Linz, Austria.

G. Cioni, Istituto di Analisi dei Sistemi ed Informatica, Consiglio Nazionale delle Ricerche (CNR), Viale Manzoni 30, I-00185 Roma, Italy.

A. Colagrossi, Dipartimento per l'Informatica e la Statistica, Presidenza del Consiglio dei Ministri, Via della Stamperia 7, I-00187 Roma, Italy.

P. Di Blasio, Dipartimento di Informatica e Automazione, Università degli Studi di Roma Tre, Via Vasca Navale 84, I-00146 Roma, Italy.

P. Forcheri, Istituto per la Matematica Applicata, Consiglio Nazionale delle Ricerche (CNR), Via De Marini 6, I-16149 Genova, Italy.

J. Gannon, Department of Computer Science, A. V. Williams Building, University of Maryland at College Park, College Park, MD 20742, U.S.A.

C. Limongelli, Dipartimento di Informatica e Automazione, Università degli Studi di Roma Tre, Via Vasca Navale 84, I-00146 Roma, Italy.

A. Micarelli, Dipartimento di Informatica e Automazione, Università degli Studi di Roma Tre, Via Vasca Navale 84, I-00146 Roma, Italy.

A. Miola, Dipartimento di Informatica e Automazione, Università degli Studi di Roma Tre, Via Vasca Navale 84, I-00146 Roma, Italy.

M. T. Molfino, Istituto per la Matematica Applicata, Consiglio Nazionale delle Ricerche (CNR), Via De Marini 6, I-16149 Genova, Italy.

I. Nemes, Research Institute for Symbolic Computation, Johannes-Kepler-University Linz, A-4040 Linz, Austria.

G. Patrizi, Istituto di Analisi dei Sistemi ed Informatica, Consiglio Nazionale delle Ricerche (CNR), Viale Manzoni 30, I-00185 Roma, Italy.

P. Paule, Research Institute for Symbolic Computation, Johannes-Kepler-University Linz, A-4040 Linz, Austria.

F. Parisi-Presicce, Dipartimento di Scienze dell'Informazione, Università degli Studi di Roma "La Sapienza", Via Salaria 113, I-00198 Roma, Italy.

A. Pierantonio, L'ACS – Dipartimento Matematica Pura ed Applicata, Università degli Studi dell'Aquila, Piazza Rivera 1, I-67100 L'Aquila, Italy.

M. Temperini, Dipartimento di Informatica e Sistemistica, Università degli Studi di Roma "La Sapienza", Via Salaria 113, I-00198 Roma, Italy.

P. Terlizzi, Dipartimento di Informatica e Sistemistica, Università degli Studi di Roma "La Sapienza", Via Salaria 113, I-00198 Roma, Italy.

State of the art and motivations

Mathematica: doing mathematics by computer?

B. Buchberger

1 Prologue

When, in the present paper, I will criticize the basic tenet of Mathematica this is not meant to detract anything from the invaluable merits of Stephen Wolfram's wholistic oeuvre with its many facets: mathematics, language design, algorithms, software and system design, teaching and publication tools, applications, and – yes – business, marketing, popularization. Rather, it is my sincere hope and wish that my analysis might help to make soon further progress in achieving, gradually, the goal of "doing mathematics by computer". Also, my proposal will show that this goal may well be achieved by suitable modifications and extensions of Mathematica and similar systems. Of course, I take Mathematica only as a paradigm for symbolic computation software systems and my analysis applies to all current systems in this category.

2 Introduction

Wolfram's (1988) manual and text book on his system Mathematica bears the subtitle "A System for Doing Mathematics by Computer". The present paper is on the question mark that may be put after this subtitle, i.e., I would like to discuss whether, indeed, Mathematica is a system for doing mathematics by computer.

For basing the argument on some solid ground, I will first discuss the following questions:

What is mathematics?
How do we "do" mathematics?
What does Mathematica "do"?

From this analysis I will draw the conclusion that Mathematica, at present, is only doing a (small) part of mathematics and I will discuss the problem of filling the gap between the existing version of Mathematica and really "doing mathematics". In my view, this problem divides into at least the following problems:

a philosophical problem,

a logical problem,
a mathematical/algorithmic problem, and
a software-technological problem.

I will discuss these problems in respective sections of this paper. Finally, I will sketch a possible extension and modification of Mathematica and similar systems that may, gradually, achieve the goal stipulated in the subtitle of Wolfram's book.

3 What is mathematics?

My personal view

There will hardly be a "definition" of mathematics that is accepted by all mathematicians. Here is my personal view that reflects what I think is essential for the analysis pursued in the present paper:

Mathematics is problem solving by reasoning.

What is a problem?

A problem is a formula P in some theory T with two free (tuples of) variables x and y (the input and output variables, respectively) that describes the properties of a desired object (denoted by) y in relation to a given object (denoted by) x.

Example: The problem of "greatest common divisor" for integers, roughly, is the formula "y divides x_1 and x_2 and, for any z that divides x_1 and x_2, $z \leq y$". Here, "x_1" and "x_2" are the input variables and "y" is the output variable.

In formulating a problem, the theory T (consisting of a bunch of constants for operations and a bunch of properties of these operations) is crucial. If the available operations in the theory T are "powerful", i.e., a lot of useful properties are known about them, the "solution" of the problem may be easy. Otherwise, finding a solution may be very difficult. In the example, T may contain the constants "divides" and "$\leq$" and, hopefully, contains some elementary properties of the corresponding operations. If T already contains the property "(t divides a and b) iff (t divides $a-b$ and b)" then a solution is "easy", otherwise a solution needs the ingenuity of Euclid, see below.

(In fact, the class of problems we are considering here is the class of "explicit problems". For the sake of simplicity, we do not consider the general class of "implicit problems" in this paper. The main point of this paper can be made clear by considering problem solving for the class of explicit problems only.)

What is a solution to a problem?

Given a theory T, a solution to a problem P (with input variable x and output variable y) is a term f in T (i.e., a term using only operations in T) with free variable x such that the following "correctness assertion for P and f" is true:

$$\forall x \; P(x, f(x)) \,,$$

i.e., "for all input values x, the value $f(x)$ has the desired property P in relation to x".

In the various languages of mathematics (including programming languages) there is a rich arsenal of constructs that allow one to build arbitrarily complex terms (programs) from elementary terms. Examples of such constructs are substitution, if-then-else, iteration, recursion, but also, in "pure mathematics", the infinite union construct, the limit construct, etc. "Algorithmic mathematics" is the branch of mathematics that is only interested in solutions that can be expressed, exclusively, by "algorithmic" constructs like substitution, if-then-else, recursion, etc. whose effects can be realized by (idealized) machines.

In a more narrow sense, algorithmic mathematics also requires that the operations of T involved in f are machine-realizable. However, the essence of algorithmic problem solving hinges on the algorithmic realization of solutions "relative" to given "black box" operations from T.

What is reasoning?

Reasoning is the activity by which new valid formulae in a theory T can be derived from the available valid formulae in the theory by using, iteratively, "inference steps" which are guided by "inference rules". In this paper, we cannot go into a deeper analysis of the notion of "inference rule" ("thinking rules", "elementary patterns for arguments", etc.). We refer the reader to any textbook on logic. The essential point for our discussion is that, whatever one's concept of "inference steps" is, *reasoning* obtains new formulae by applying inference rules and, thus, is in sharp contrast to the activity of *observation*, which is used in natural sciences like physics for obtaining new valid formulae. Whereas *observation* is only possible by bringing the senses into contact with the world described by a formula, *reasoning* is an activity that essentially does not involve the senses but the mind, which operates on the level of the linguistic model of the world or, stated differently, on the level of the formulae itself. Whereas *observation* allows one to obtain new knowledge only about individual facts, *reasoning* allows one to derive universally quantified, "general" formulae (from other universally quantified formulae). Also, whereas *observation* allows one to obtain "absolute" knowledge, *reasoning* derives knowledge whose validity is only relative to the validity of the knowledge from which reasoning started.

Observations should be "objectively verifiable" in the sense that "in the same situation everybody should be able to observe the same fact". Reasoning should be "objectively verifiable" in the sense that "everybody should be able to check that the individual inference rules were appropriately applied". Objective observation and objective reasoning together form the two pillars of the "rational" or "Western" or "mind-oriented" approach to science.

For appreciating the rest of this paper and the practical implications for the analysis of whether or not Mathematica is a system for doing mathematics by computer, it is not necessary that you agree with the above oversimplified and rough analysis of reasoning and observation. It suffices that you agree with me that there is a fundamental difference between reasoning and observation, and, more importantly, that it is the essence of mathematics that new knowledge is obtained by reasoning or, more technically, that *the essence of mathematics is proving* or, still more explicitly, that "mathematics" without proving is not really mathematics.

In the context of problem solving, hence, a *mathematical* solution f to a problem P needs a proof of the corresponding correctness assertion for P and f. Note that this view is independent of the particular notion of proof an individual mathematician has. In particular, if we speak about "proof" here, we do not necessarily mean "formal" proof, i.e., a proof whose individual inference steps are governed by machine-realizable inference rules. Personally, however, I am inclined to believe that it is an excellent training and, in fact, a practically feasible goal to work out proofs in a way that would easily allow breaking them down into machine-realizable inference steps. What is important is that the crucial assertions from which the correctness assertion for P and f can be obtained must be *proved.* In order that f is a *mathematical* solution to problem P it is not sufficient that these assertions are *"observed"* in a few (or even many) instances like assertions in experimental physics.

In the example of the "greatest common divisor" problem the following solution term

$$\text{“}x_1 > x_2 \Rightarrow \mathrm{GCD}(x_1, x_2) = \mathrm{GCD}(x_1 - x_2, x_2)\ ,$$
$$x_1 < x_2 \Rightarrow \ldots\text{”}\ ,$$

which is called "Euclid's algorithm", at least needs a proof of the crucial theorem on which this algorithm hinges:

$$\forall t\ ((t|a \wedge t|b) \iff (t|a - b \wedge t|b))\ .$$

(Here we wrote "$t|a$" for "t divides a".) I hope all mathematicians, independent of their view of what *is* a proof, will agree that this theorem needs a proof or, otherwise, Euclid's algorithm would not be accepted as a *mathematical* solution to the problem of greatest common divisor. Experimental verification of the "theorem", even in a great number of instances, would not be sufficient.

Of course, it is well known that Euclid did *not* give a proof of the above theorem but, in fact, only demonstrated "plausibility" in a few instances. However, this historical fact just demonstrates that, at the dawn of mathematics, the border between "physics" (observation) and "mathematics" (proving) *was* vague and it also illustrates the real difference between modern full-blown mathematical thinking and ancient first steps into mathematics. Stated differently, one could say that, if one believes that mathematics does not necessarily mean proving, one turns the wheel two thousand years back. Two thousand years of progress of mathematics essentially was progress in proving techniques.

4 How do we "do" mathematics?

Some, in particular those who do "experimental mathematics", may be reluctant to agree with this "clean" view of mathematical problem solving that emphasizes proving. In fact, what I said so far was only meant to characterize what mathematics *is*. It is a different story if we start to analyze how we *do* mathematics, i.e., *how we get ideas* for interesting new knowledge that is reasonably conjectured to be true and that may be useful for solving a given problem and *how we find the inference steps* of a proof that eventually establishes the universal validity of a conjectured formula.

Here is my personal view of how I believe that mathematics is "done":

Mathematics proceeds by iteration through the "creativity spiral".

The concept of *spiral* has two ingredients: on the one hand, one proceeds through a *circle* and, on the other hand, one arrives at *a higher level*. Arriving at the next level in a spiral saves one from becoming "circular", i.e., proceeding indefinitely without making progress. Given a problem, one "circle" of the creativity spiral has the following four essential steps:

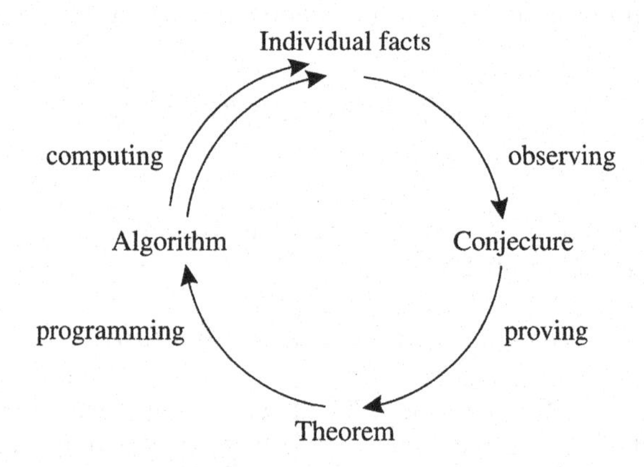

According to the circular nature of the creativity spiral, it is completely arbitrary that I start the description of one circle in this creativity spiral with an "algorithm" for a given problem.

1. For example, for the problem of greatest common divisor the most primitive *algorithm* is nothing else than a procedural reading of the definition of "greatest common divisor". It consists of trying out all divisors t of the given x_1 and x_2, marking the common divisors of x_1 and x_2 and saving the greatest one. *Computing* the greatest common divisors of many pairs (x_1, x_2) of numbers one may collect sufficiently many facts about individual instances of the problem (including the "traces" of the computations) to be ready for the next step in the creativity spiral.

2. By *observing* the common structure "hidden" in the experimental data, one may obtain the insight that the problem of determining the greatest common divisor of a and b (where $a > b$) can be reduced to the problem of determining the greatest common divisor of $a - b$ and b. Note that this is truly an *in-sight* in the sense that it is obtained "by observing and seeing", i.e., an action of the

senses. This is not yet the place for reasoning. An insight, formulated as an assertion, is a *conjecture*. (In the particular example considered here, the insight might also arise from other "heuristic activities", for example from analyzing the logical structure of the formula that defines the notion of "greatest common divisor".)

3. A conjecture, however, is not yet a *theorem*. A theorem needs a *proof*, i.e., a sequence of thinking steps that shows that the conjectured assertion is not only true in the cases observed in the computational experiments but is necessarily true in all possible cases. In our case, this proof may roughly run as follows: "If (an arbitrary) t divides (arbitrary) a and b then there exist u and v such that $t \cdot u = a$ and $t \cdot v = b$. Hence, $a - b = t \cdot u - t \cdot v = t \cdot (u - v)$, which shows that t also divides $a - b$. Conversely, if ..." The individual thinking steps do not depend on special properties of the few instances of numbers studied in the experiments but cover the "general" case or certain general subcases. It needed the intellectual effort of many mathematicians over the last two thousand years to develop thinking techniques that "guarantee" that proofs formulated in a finite amount of text can derive true statements about infinitely many objects. This is the "miracle" of proofs in mathematics. Philosophically, this could be discussed endlessly. However, for the purpose of this paper, it should be clear that this *is* the present scenario of mathematics.

In the phase of *proving*, again, the insight obtained in the stage of observing and studying the individual experimental facts obtained in the computational experiments may be crucial for discovering the sequence of thinking steps needed in order to obtain a complete proof. Of course, in some simple cases, as in our example, no great insight is necessary and an experienced "prover" may obtain most of the proof just by "applying the definitions" of the concepts involved, for example the definition of "$t|a$". In the proof of "nontrivial" theorems, however, a great deal of "insight" is necessary for "seeing" why the theorem is true, i.e., to find the crucial nontrivial steps in the proof that go beyond mere routine application of definitions and other routine proving techniques.

4. A "good" theorem has problem solving power, i.e., it shows how the problem considered can be solved or, if a "nonalgorithmic" solution is already known, how it can be solved in an "algorithmic" way or, if an algorithmic solution is already known, how it can be solved in a more "efficient" way, i.e., with less effort in terms of computational time, memory, etc. Basically the problem solving power of a theorem consists in the description of how the problem at hand can be reduced to other, hopefully simpler, problems. The explicit and detailed description of this reduction is the "method" for solving the problem that can be extracted from the theorem. When this reduction can be described by language constructs that can be realized by machines then we, hence, can obtain an *algorithm* from a theorem. The extraction of algorithms, formulated in a particular algorithmic language, from theorems is also called *programming*.

Now, one circle in the creativity spiral is closed and we are, again, in the position where we can apply the algorithm for particular problem instances, collect more experimental facts and, hopefully, get new insight. Note, however, that the next pass through the circle can now proceed on a *higher level*: Since the algorithm obtained incorporates more insight, and hence is more efficient, than the algorithm available at the beginning of the first round, we may consider more complicated examples in less time and the experimental data may be much richer and may lead to much more interesting insight, which again may give rise to a deeper theorem with a deeper proof and finally may result in an even more efficient algorithm. In our example, the consideration of more complicated pairs of integers, with hundreds of digits, may lead to the observation that pairs sharing the same initial digits, for a long period in the trace of Euclid's algorithm, will have similar sequences of intermediate remainders. This will eventually lead to Lehmer's improved version of Euclid's algorithm.

By considering the creativity spiral of mathematics, one also gets a clear understanding of the fundamental role of mathematical knowledge incorporated in theorems: Obtaining mathematical knowledge is not (so much) a matter of aesthetics or enjoying a static world of beautiful structure but it is the indispensable foundation of solving problems and, if possible, solving them mechanically and efficiently. Interestingly, although mind-oriented "Western" science seems to be 180 degrees opposite to consciousness-oriented "Eastern" science, problem-orientation is shared by both approaches to science. Thus, in the Bhagavad-Gita, one of the most fundamental sources of consciousness-oriented science based on meditation, the final objective of "Yoga" (science based on consciousness) is described as "Firmly established in insight, act!".

5 What does Mathematica "do"?

In terms of the steps of the creativity spiral of mathematics, I think that Mathematica does the following:

1. Mathematica is a (very convenient) programming language, i.e., it allows one to formulate and execute algorithms. It does better in this respect than many other programming languages, in particular because it also allows one to program in a "rewrite rule style", which is so close to conventional mathematical language that sometimes one may take theorems nearly in the form as they appear in mathematical papers and almost immediately can use them as Mathematica algorithms. (Buchberger (1996) describes how, in the future, one will be able to take the theorems *exactly* in the form they appear in mathematical papers and use them directly as algorithms.) Also, Mathematica comes equipped with a huge collection of built-in algorithms for solving problems in many areas of mathematics, which can be used as building blocks for more complex new algorithms. Hence, Mathematica is an excellent tool for the "programming" step in the creativity spiral.

2. Mathematica, in addition, provides very advanced facilities for graphical representation of data, tracing algorithms, and experimenting with algorithms while reading a mathematical text (the notebook facility). Thus, Mathematica is also an excellent tool for the "computing" and the "observing" phase of the creativity spiral, where we want to collect interesting experimental facts and, by convenient manipulation of both the algorithms and the data, would like to obtain new insight.

3. Mathematica has also a limited capability of "proving" because the "sim-

plifier" that forms the central kernel of Mathematica incorporates many of the rewrite rules that are normally used in mathematics for transforming "expressions". For example, expressions can be expanded: When we enter

$$\mathrm{Expand}[x(y+z)]$$

the system produces: $xy + xz$.

Similarly, the equivalence of expressions can be tested. With the input

$$\mathrm{TrueQ}[\mathrm{Expand}[n(n+1)/2+n+1] = \mathrm{Expand}[(n+1)(n+2)/2]]$$

the system produces: True.

The latter evaluation automatically verifies the crucial step in the inductive proof of Gauss's formula for the sum of the first n natural numbers. In fact, it is just the essence of symbolic computation systems that many of the equivalence transformations and equivalence tests that mathematicians apply in order to verify "formulae" or to derive new ones can be executed on machines or can, at least, be supported by machines. In the case of Mathematica, the collection of these transformations also establishes the interpreter for Mathematica as a programming language.

6 The gap between mathematics and Mathematica

The arsenal of transformations supported by Mathematica-like systems is quite impressive and certainly increases the problem solving potential of mathematicians by an "order of magnitude" (which is a way of defining something nobody can really define). However, these rewriting transformations still cover only a very small part of what "proving" really is. Here are the additional ingredients that go beyond application of rewrite rules in proving:

- *(Use definitions):* Replace defined concepts by the defining formula, which may involve quantifiers.
- *(Prove "for all"):* Eliminate universal quantifier by the "arbitrary but fixed" technique.
- *(Use "for all"):* Use universally quantified formula for special instances.
- *(Prove "there exists"):* Eliminate existential quantifier by inventing a "solving term".
- *(Use "there exists"):* Use existentially quantified formula by the "let ... be such a ..." technique.
- (Other, in particular, *domain-specific quantifiers*, like set braces, limes etc.): Eliminate these quantifiers or use such quantified formulae.
- *(Find rewrite sequence):* Find the appropriate sequence of rewrite steps for establishing equivalence (or other transitive properties) of expressions.
- Various forms of *induction* ...
- Many *domain-specific proof techniques* that could, in principle, be replaced by universal techniques w.r.t. a set of axioms whose availability, however, is very important for making proving practically feasible.

Some of these ingredients of "real" proving are mere routine and their implementation in a system for "doing" mathematics would be both easy and helpful. Others, notably (prove "there exists"), (find rewrite sequence), some of the rules for (domain-specific quantifiers), and many domain specific proof techniques need "ingenuity". However, as can be shown by inspecting many examples of practical proofs, interesting theorems, after hierarchical decomposition by lemmas, normally do not involve more than one, two, or very few "ingenious" steps, which we well should and could leave to human interaction also in the near future. The rest is mostly quite routine but tedious. For example, when the theorem that lies behind Euclid's algorithm is already conjectured, its proof does not involve any ingenious steps. It proceeds by using definitions, handling quantifiers and some rewriting steps. The only not completely trivial step is to see that $(u - v)$ is the term suitable for showing that t divides $a - b$. In preparation to this, some skillful concatenation of rewrite steps (transforming $t \cdot u - t \cdot v$ into $t \cdot (u - v)$) is necessary).

Still, the routine steps and book-keeping mechanisms necessary for doing real proofs are far beyond the capabilities of Mathematica and similar systems. Therefore, I do not think that it is yet appropriate to say that Mathematica is a system for "doing" mathematics by computer. However, I *do* believe that it would well be possible to create systems that include all the potential of Mathematica and similar systems but, except for the (few but important) nontrivial proof situations that need ingenuity, also give significant assistance to real proving. For achieving this goal, several problems have to be overcome, which I will discuss briefly in the sequel.

7 A philosophical problem

Before one can reasonably start to pursue the goal of incorporating proving into a system that "does" mathematics by computer one must, trivially, be convinced that proving is a crucial, if not *the* crucial, ingredient of mathematics. I am amazed that apparently the designer of Mathematica is amazed about the fundamental role of proof in mathematics:

"I continue to be amazed that, in the math educational process, there is such an emphasis on, as I see it, highly esoteric issues of pure mathematics. The notion of proof is an interesting one, but very few people in adult life so to speak, 'do proofs'. That kind of thinking is, I think, most prevalent among mathematicians and lawyers. I think the emphasis on that kind of thing in mathematics education is a consequence of some kind of trickle-down effect from the influence of mathematics research in this century" (Wolfram 1993).

Although I like most aspects of Wolfram's work including most of his "political" statements about science, business, software, etc., I just cannot agree with his view on proving and its role in "doing" mathematics.

As explained in my model of the "creativity spiral of mathematics", proving *is* an essential part of mathematics both by the position and role of mathematics within science and by the actual practice of mathematics.

Pure and *algorithmic* mathematics cannot be distinguished by the observation that pure mathematics involves proving while algorithmic mathematics is experimental and can get along without proofs. In my view, pure (or "non-algorithmic") and algorithmic mathematics are distinguished by the fact that pure mathematics allows a broad class of operators for constructing new solution terms from simpler terms while algorithmic mathematics allows "machine-realizable" operators like substitution, if-then-else, iteration, etc. only.

Both pure and algorithmic mathematics need proofs. In fact, *algorithmic mathematics most times needs more sophisticated theorems and more difficult proofs* than "pure" mathematics. This is so because the solution of problems becomes harder when the collection of available operators is restricted to "machine-realizable" ones, i.e., when the collection of available operators becomes smaller.

Politically, Wolfram's argument is ridiculous because some of the strongest parts, and actually the nontrivial parts, of his system rely on the work of mathematicians who have diligently spent their time and ingenuity for proving results on which the algorithms of Mathematica and similar systems are now based.

Thus, frankly, I believe that the designer of a system for "doing" mathematics by computer will have to appreciate the fundamental role of proof in mathematics before a system that meets this tenet can be produced.

8 A logical problem

From now on, let's assume that we really would like to create a system for doing mathematics by computer, i.e., a system that also supports proving. A natural approach would ask for just extending Mathematica. I think that, in principle, this is a feasible undertaking. However, some care must be taken because there are quite a few differences in the basic design of "mathematics" and Mathematica.

Embedding algorithms into predicate logic

A natural system for doing mathematics by computer would want to start from the language frame in which these days most of mathematics is written (or could be written after some cleaning up of the formal details), namely a version of predicate logic, into which set theory is embedded by including the set theoretical symbols and set theoretical knowledge. Algorithms, then, are just special sequences of formulae in this system namely, essentially, sequences of equalities. Sequences of equalities in this setting form a programming language whose interpreter is the rewrite rules part of predicate logic.

For example, in the case of Euclid's algorithm we would like to formulate, in one language frame, the definition of "divides", the definition of "greatest common divisor", the main theorem on which the algorithm is based, and the algorithm itself:

Definition of "divides":

$$t|a \iff \exists u(t \cdot u = a) .$$

Definition of "greatest common divisor":

$$\mathrm{GCD}(a, b) = \max_t (t|a \wedge t|b) .$$

Main Theorem:

$$t|a \wedge t|b \iff t|(a-b) \wedge t|b .$$

Euclids' Algorithm:

$$a > b \Rightarrow \mathrm{GCD}(a, b) = \mathrm{GCD}(a-b, b) ,$$
$$a < b \Rightarrow \ldots .$$

Of course, one could easily extend the syntax of Mathematica for including various quantifiers. Also, the differences in the syntax between usual mathematics and Mathematica is not a real issue. In fact, the new version of Mathematica will have sophisticated type-setting facilities that can be used both in text cells and in input cells and, thus, the gap between the syntax of every-day mathematics and Mathematica will be closed almost completely. (There will remain some syntactical differences which are nasty, e.g., the use of brackets for function application and the use of the braces for tuples instead of set.)

However, as it stands now, Mathematica is only a programming language. That means that formulae involving higher constructs from logic, e.g., formulae with quantifiers, cannot be processed by the system. Rather, they can only be considered as part of the explanatory text in the "text cells". This is in sharp contrast to the Mathematica formulae (directed equalities) in the "input cells", which have logical, at least computational, meaning.

Speaking about the differences in the design of mathematics and Mathematica, one must, however, also see that the present "Bourbakistic" logical frame for mathematics (i.e., first-order predicate logic plus set theory) is not an appropriate frame for embracing both the algorithmic and the nonalgorithmic aspect of mathematics. The point is that the "extensionality axiom" of set theory identifies functions whose input/output behavior is identical even if the definition of the function (e.g., by an algorithm) is different. This does not allow us to speak about properties of functions that depend on their definition, e.g., complexity issues. Thus, in order to close the gap between the design of "mathematics" and the design of systems like Mathematica, one must also drastically change the design of present-day mathematics. A natural way out is the use of a suitable variant of higher order predicate logic. We give more details about this subtle point elsewhere (Buchberger 1996).

Equational logic and Mathematica's simplifier

It is a remarkable, though little known fact that the "simplifier" of Mathematica that forms the innermost kernel of the system can be viewed as an implementation of the (directed) equational part of higher order predicate logic. In my opinion, this makes Mathematica superior to most other symbolic computation systems. In fact, this feature makes Mathematica a natural candidate for a further expansion of the logical power of symbolic computation systems along the lines proposed in the present paper.

Unfortunately, the fact that the kernel of Mathematica is an implementation of higher order equational reasoning is hidden by the design decision that, in Mathematica, all variables implicitly range over the complex numbers or a subdomain of the complex numbers, i.e., the general simplifier inherits certain special equational properties of the complex numbers, e.g., commutativity, associativity.

For being a frame for all of mathematics, a future system should have its programming language defined essentially as the equational logic part of predicate logic based mathematics in order to avoid incompatibilities. Only if deliberately chosen, a special version of equational logic, taking into account additional equational axioms like commutativity etc., should be invoked.

A variety of simplifiers

On top of a general higher order equational logic "simplifier", a future mathematics system should have a variety of other "simplifiers" or "provers". Each of these simplifiers would be appropriate for a different class of formulae and for a certain selection of the formulae currently active in the text. In the above example of the Euclidean algorithm and its underlying "theory", we would need a "natural deduction" prover for the actual proof of the main theorem but we need a left-to-right directed equational prover or "interpreter" that is applied selectively only to the formulae that constitute Euclid's algorithm when we want to "compute" the GCD for an individual input. Thus, in a system for "doing" mathematics it is not sufficient to have just one general simplifier that is applied in all situations.

Here is a list of simplifiers (provers) that should be available for appropriate invocation in the various situations:

- A general interactive "natural deduction" prover that organizes the overall proof procedure for formulae in arbitrary theories formulated in the frame of predicate logic. Essentially, this prover has to handle quantifiers and organize the book-keeping process for sets of formulae with nested quantifiers. It should be modeled exactly after the routine followed by "human" provers, see Buchberger and Lichtenberger (1980).
- A general equational theorem prover.
- A fully automated propositional logic prover for the (rare) situations in real proofs, where the proof situation can be handled by propositional logic only.
- Special rules embedded into the natural deduction prover for handling the set theoretic quantifiers.
- Special rules for handling quantifiers from other theories, e.g., the quantifiers of analysis.
- Special versions of the equational theorem prover in the presence of equational axioms or derived properties for some of the functions, e.g., the arithmetical operations or the elementary transcendental functions.
- Special versions of the equational theorem prover using advanced techniques like the Knuth–Bendix procedure, narrowing, etc.

- Directed equational theorem proving (in the presence of axioms) as the "interpreter" for the rewrite-type programming language embedded into predicate logic and a procedural language defined in terms of it.
- Theorem provers based on algebraic techniques like Gröbner bases, characteristic sets, etc. for certain special classes of geometrical theorems.
- Theorem provers for the theory of real closed fields, e.g., Collins's prover, and theorem provers for other special theories.
- A resolution theorem prover for sets of formulae involving quantifiers in some special situations of low complexity, where the invocation of resolution promises to give a result in reasonable time.
- Induction provers for inductively defined domains.

Summarizing, a variety of "general" provers applicable in almost all domains, and special provers for each particular domain, should be made available in future systems. The interplay of the special provers is a nontrivial problem that could be solved by building up mathematics using a hierarchy of "functors". For more details about this approach and a sketch of a realization within Mathematica, see Buchberger (1996).

9 A mathematical/algorithmic problem

I think that the mathematical and algorithmic problems connected with including various different provers and simplifiers into one overall system for doing mathematics by computer are basically solved. There are many simplifiers and provers around for special classes of formulae and their respective scope of "tractable" sets of formulae is well understood. Also, arranging proofs in the natural deduction style is well advanced.

Thus, the main problem consists in putting the pieces together and forming a uniform system that gives the user convenient control over the entire process of doing mathematics within the system. So, one main question is that "somebody" decides to invest the time and money into the software-technological and -commercial problem of developing and marketing such a system. The other main question is how the interplay between the various complex pieces can be broken down in a manageable way. Here, again, I think that functors play a crucial and, in fact, new role, see Buchberger (1996).

10 A software-technological problem

The software-technological problems involved in creating a system for "doing" mathematics in the broad sense of the creativity spiral seem to be manageable. In fact, the very existence of Mathematica (more than any other system) shows that all the software-technological tools for efficient implementation of formulae "simplifiers" and provers, man–machine interaction, graphical computing, interactive teaching and studying using the notebook facility, professional mathematical typing, etc. are available. What is really necessary is the integration of the higher forms of logical reasoning into one common system frame.

11 Two examples

Here are two examples that sketch how doing mathematics in a *Mathematics* (instead of Mathematica) system could look like. (Input is typed roman and output is typed italic. The syntax contains much "sugar", which is inessential, though. The command ENTER has a parameter that indicates which "simplifier" or "prover" should be used in order to process the formulae entered.) The natural deduction theorem prover used in the first example is partially implemented and the induction prover used in the second example is completely implemented by the author using Mathematica as an implementation language.

11.1 Natural deduction for a proof on divisibility

ENTER into knowledge base:

(Definition of "divides"):

$$t|a :\Leftrightarrow \exists u(u \cdot t = a) \ .$$

(Definition of "greatest common divisor"):

$$\mathrm{GCD}(a, b) := \max_{t}(t|a \wedge t|b) \ .$$

(Order lemma for "divides"):

$$t|a \Rightarrow t < a \ .$$

Echo the formulae . . .

ENTER into natural deduction prover:

(Bounded "GCD"):

$$\mathrm{GCD}(a, b) = \max_{t<a}(t|a \wedge t|b) \ .$$

True by (definition of GCD) and (order lemma for "divides").

ENTER into algorithm interpreter using (bounded GCD):

GCD(18, 12)

6

GCD(24, 18)

6

.... (more experiments)

ENTER into natural deduction prover:

(Conjecture about GCD):

$$\mathrm{GCD}(a, b) = \mathrm{GCD}(a - b, b) \ .$$

By (definition of GCD) we have to show

$$\max_t(t|a \wedge t|b) = \max_t(t|a-b \wedge t|b) \ .$$

It suffices to show that

$$t|a \wedge t|b \iff t|a-b \wedge t|b \ .$$

Let t be arbitrary but fixed and assume

$$t|a \wedge t|b \ . \qquad (D0)$$

We have to show

$$t|a-b \qquad (D1)$$

and

$$t|b \ . \qquad (D2)$$

(D2) follows propositionally. For (D1), by (definition of "divides"), we have to find a w such that

$$t \cdot w = a - b \ . \qquad (D1')$$

From (D0), by (definition of "divides"), we know that for certain u and v,

$$t \cdot u = a$$

and

$$t \cdot v = b \ .$$

Hence,

$$a - b = t \cdot u - t \cdot v \ .$$

Question: Which w in (D1')?

ENTER into integer simplifier:

$$t \cdot u - t \cdot v$$
$$t \cdot (u - v)$$

ENTER into natural deduction prover in (D1′):

$$w = u - v$$

True.

And so on.

ENTER into algorithmic interpreter using (conjecture about GCD):

GCD(3345, 1230)

15

And so on. Of course, tedious "traces" of proofs need not be printed out. Rather, in the case of the above proof the ouput could be reduced to:

The proof reduces to:

$$\textit{Find } w \textit{ such that } t \cdot u - t \cdot v = t \cdot w \ .$$

Alternatively, the output could be just *True* if we have a suitable unification algorithm (for commutative-associative-distributive function symbols) built-in so that user interaction for finding a suitable w can be replaced by algorithmic unification.

11.2 Induction for elementary properties of operations on the natural numbers

We show one example of the result of applying a "multiple induction prover" for formulae ranging over the natural numbers. The prover is written in Mathematica (using substitution as the most important construct) and can be applied to universally quantified equalities presented as Mathematica expressions. Here, "multiple" refers to the fact that equational formulae with several universal quantifiers can be handled. The induction prover is called recursively. The prover first tries to prove the formula by simplification without induction (taking all universally quantified variables as being "arbitrary but fixed constants"). If this is not possible it undertakes induction with respect to the first universally quantified variable. If the induction base or the induction step cannot be proved by simplification it attempts an induction proof with respect to the second universally quantified variable and so on.

It turns out that this simple strategy is quite powerful and allows to prove complex equational theorems with very little help from the user. We have generated quite a few such proofs, which show several interesting facts:

- The proof texts generated by the prover are self-explanatory. Some of the texts might be shortened for experienced readers. In fact, the text can be modified by setting several option variables.
- The prover may well travel through quite a few levels of induction following the natural sequence of universally quantified variables and finally arrive at a successful proof.
- The complexity of the proofs depends heavily on the sequence of the universally quantified variables. We did not yet build in (known) heuristics for determining an optimal sequence for these variables.
- One sees that developing a mathematical theory in the presence of a powerful mathematical proving assistant amounts to inventing conjectures in a suitable order so that the prover can master the individual proofs. Proofs that fail may give a hint for a conjecture that may be handled in the next round of the prover: The formula left over when all nested inductions have been tried out and simplification does not reduce both sides of the equality to identical

terms is normally a good candidate for a lemma to be proven before going back to the proof of the main theorem.

In this example, we consider the theory T (the additive theory of natural numbers) that consists of the following two axioms:

$$x + 0 = x \ ,$$
$$x + y' = (x + y)' \ ,$$

which, for our prover, must be encoded as Mathematica rules in the following way:

$$T = \{\, \mathrm{sum}[x\_, 0] :> x,\ \ \mathrm{sum}[x\_, s[y\_]] :> s[\mathrm{sum}[x, y]] \,\} \ .$$

The multiple induction prover can then be called in the following way for proving, e.g., associativity:

ENTER into multiple induction prover for theory T:

$$\forall z, y, x (x + (y + z) = (x + y) + z) \ .$$

The following output is then produced completely automatically:

Proof:

Try with arbitrary but fixed z0, y0, x0 simplification proof of:

. $((x0 + (y0 + z0)) = ((x0 + y0) + z0))$.

.. *Simplification of left-hand side:*
.. $(x0 + (y0 + z0))$

.. *Simplification of right-hand side:*
.. $((x0 + y0) + z0)$
. *Not proved by simplification.*

Induction proof of:

$\forall z, y, x\ ((x + (y + z)) = ((x + y) + z))$

Induction variable: z.

. *Prove induction basis (i.e. formula with* $z \to 0$*):*

. $\forall y, x\ ((x + (y + 0)) = ((x + y) + 0))$

. *Try with arbitrary but fixed y0, x0 simplification proof of:*

. $((x0 + (y0 + 0)) = (x0 + y0) + 0))$.

... *Simplification of left-hand side:*
... $(x0 + (y0 + 0))$
... $= (x0 + y0)$

... *Simplification of right-hand side:*
... $((x0 + y0) + 0)$
... $= (x0 + y0)$
.. *Proved by simplification.*
. *Proved induction basis with* $z \to 0$.

. *Let z0 be arbitrary but fixed.*
. *Induction hypothesis (i.e. formula with* $z \to z0$*):*

. $\forall y, x\ ((x + (y + z0)) = ((x + y) + z0))$

. *Prove induction step formula (i.e. formula with* $z \to z0'$*):*

. $\forall y, x\ ((x + (y + z0')) = ((x + y) + z0'))$

. *Try with arbitrary but fixed y0, x0 simplification proof of:*
. $((x0 + (y0 + z0')) = ((x0 + y0) + z0'))$.

... *Simplification of left-hand side:*
... $(x0 + (y0 + z0'))$
... $= (x0 + (y0 + z0)')$
... $= (x0 + (y0 + z0))'$
... $= ((x0 + y0) + z0)'$

... *Simplification of right-hand side:*
... $((x0 + y0) + z0')$
... $= ((x0 + y0) + z0)'$
.. *Proved by simplification.*
. *Proved induction step formula with* $z \to z0'$.
Proved by induction.

Note that if we chose the universally bound variables in the order x, y, z then the proof of the induction basis and the induction step of the induction over x by simplification of the left-hand and right-hand sides of the resulting equalities would not work. Rather, both in the base case and also in the induction step for x, the prover starts induction over y automatically and so on. Finally, it finds a completely automatic proof also for this order of variables, albeit an unnecessarily complicated one.

12 Conclusion

We discussed the question of whether present symbolic computation systems can truly be considered as "doing mathematics by computer". Although our answer was "no" for the present systems we believe that, with research directed into appropriate directions, significant progress will be possible in the near future for making symbolic computation systems assisting humans more fundamentally when "doing mathematics". Various research projects aiming at this goal are under way worldwide. Partly, they start from computer algebra systems like Mathematica and move into the direction of incorporating more proving power. Partly, they start from sophisticated provers and move into the direction of incorporating more algebraic computation power. The author is currently working along the first of the two strategies.

Acknowledgements

The author's research on the topics described is carried out in the frame of a joint project with Fujitsu Labs, Numazu, Japan.

References

Buchberger, B., Lichtenberger, F. (1980): Mathematik für Informatiker I: Die Methode der Mathematik. Springer, Berlin Heidelberg New York.

Buchberger, B. (1996): Symbolic computation: computer algebra and logic. In: Baader, F., Schulz, K. (eds.): Frontiers of combining systems. Kluwer, Dordrecht.

Wolfram, S. (1988): Mathematica: a system for doing mathematics by computer. Addison-Wesley, Redwood.

Wolfram, S. (1993): An interview with Stephen Wolfram. Mathematica Educ. 2/2.

An overview of the TASSO project

A. Miola

1 Introduction and motivations

Symbolic and algebraic computation systems (Derive, Macsyma, Maple, Mathematica, Reduce, to name a few) have been made widely available since several years. These systems have supported the mathematical problem solving in several application areas of sciences and engineering with significant achievements (Buchberger et al. 1983; Pavelle 1985; Caviness 1986; Davenport et al. 1988; Miola 1990a, b, 1993b, a).

However those systems are still far from being completely correct and safe from the qualitative point of view of a computation. Let us address some of the problems which can rise in using the currently available systems. We do not refer here to any system in particular. As a matter of facts the problems we like to consider here have to be interpreted as general and common problems in symbolic computation, when using a system designed and implemented with an inadequate methodological approach.

Let us discuss this matter through very simple examples which already lead to unsatisfactory computing situations.

Example 1.
1. $(\sqrt{x})^2 \rightarrow x.$
2. $\sqrt{x^2} \rightarrow x.$

Example 2.
Step 1: Let y be $\frac{1}{x}$,
Step 2: let z be $\frac{\mathrm{d}}{\mathrm{d}x} \log x$,
Step 3: $y - z \rightarrow 0$.

Examples 1 and 2 are mathematically incorrect. The function $\frac{1}{x}$ and the function $\frac{\mathrm{d}}{\mathrm{d}x} \log x$, even with the same analytical expression $\frac{1}{x}$, are defined in two different domains. The functions $(\sqrt{x})^2$ and $\sqrt{x^2}$ are different, as being defined in two different domains. And this is not taken in any consideration by a system designed simply on the basis of syntactic rewriting rules.

Example 3.
Step 1: Let y be x^i,

Step 2: $\sum_{0\le i\le 10} y \to 11x^i$.

Example 4.
Step 1: Let y be $\sum_{i=1}^{n} x_i$,
Step 2: $\frac{\partial y}{\partial x_1} \to 0$.

Examples 3 and 4 are wrong for completely different reasons. The symbol i is used with different meanings in the two successive steps. A system designed without appropriate semantics can often lead to such unpleasant situations.

Example 5.
1. $\infty - \infty \to 0$.
2. $\infty + \infty \to 2\infty$.
3. $\infty \cdot 0 \to 0$.

Example 5 shows the obvious need for a precise algebra and therefore for an adequate semantics to deal with infinity quantities.

In general, a message like "division by zero" occurs in case of Example 6.

Example 6.

$$\sum_{-1\le i\le 10} \int x^i \, \mathrm{d}x \to ?\ .$$

This example again presents the case where the symbol i of the integrand function x^i is not bound by the constraint given in the summation. In order to get rid of this, the computation could proceed as follows, by interchanging the sum and the integral operators and splitting the sum into parts:

Step 1: $\to \int \sum_{-1\le i\le 10} x^i \, \mathrm{d}x$,
Step 2: $\to \int (x^{-1} + \sum_{0\le i\le 10} x^i) \, \mathrm{d}x$,
Step 3: $\to \log x + x + \frac{x^2}{2} + \ldots + \frac{x^{11}}{11}$.

Certainly some mathematical knowledge is needed to carry out this computation, then, a powerful and flexible system is needed to support this kind of computation.

Starting from these motivations the research project TASSO is conducted to study and define an appropriate methodology to design and implement symbolic computation systems. This project is mainly based on specification and programming methodologies for axiomatizable objects manipulation.

2 TASSO structure

TASSO deals with abstract mathematical entities as objects described by axiomatic specification. It considers logic formulas and algebraic structures. Each object has a unique formal definition with the specification of its attributes and algebraic, analytical and logic computations are possible under fixed constraints. In particular, the project is based on the following aspects:

- programming methodologies for abstract specifications;
- axiomatic definitions of mathematical objects;
- automated deduction mechanisms, as a new basic computing tool;
- algebraic and heuristic methods for applied mathematics, at a high level of abstraction with respect to the domains where the problems are defined.

The specification of an object is given by following the typical object-oriented mechanism of inheritance, so as to reproduce the classical hierarchy of the mathematical abstract structures. For instance, the abstract algebraic structure *ring* is specified by assuming the specification given for the structure *group*, as in abstract algebra.

The dynamic definition and manipulation of mathematical objects is also possible: the specification of an object (e.g., matrix, polynomial) is unique and the instantiation mechanism allows one to compute with objects defined at run time (e.g., matrix over polynomials over . . . matrices of integers).

Moreover, the specification of an object includes the specification of admissible computing methods, given through an abstract definition at the highest possible level of the object hierarchy. For instance, the *Euclidean algorithm* is specified as an attribute of the object *Euclidean domain*. Then the instantiation of the Euclidean domain implies the instantiation of all the related computing methods (e.g., the Euclidean algorithm).

From an application point of view both numerical and non-numerical computing methods are needed in a software system for symbolic computation. Actually, the introduction of numerical methods can be done consistently with the methodological approach of the entire project, so as to offer the two types of computing methods in a single integrated computing environment.

In order to develop such an integrated environment, one starts from the well known similarity of numbers and polynomials, with their related arithmetic operations, and builds very general abstract data structures to encapsulate various similar objects. At the same time, the algebraic p-adic construction (Hensel method) and the classical numerical approximation method (Newton method) can be viewed as special cases of a more general algebraic approximation method in abstract structures.

Then, the available specification mechanism allows one to exploit the main results on the integration and the *amalgamation* of numerical and algebraic methods (Limongelli et al. 1990; Miola 1990a, b; Limongelli and Miola 1990; Limongelli and Temperini 1992). The concept of amalgamation is borrowed from logic programming and expresses the methodological approach to software development based on two levels of operations specified with uniform semantics. The operations of the upper level control the execution of the operations at the lower one. Actually, numerical computing can be interpreted as a lower level where the computations are completed under the control of symbolic computing at the upper level, where the complete formalization of the entire computation is given.

Furthermore, the result on the possible unified interpretation of the Hensel construction for algebraic equation solving and of the Buchberger method for the solution of systems of polynomial equations could support this definition of a general method for abstract approximation construction (Miola and Mora 1988).

The inheritance mechanism in specifications is strictly related to the notion of subtyping, as any derivation obtained by inheritance corresponds to a compatible assignment rule. Furthermore, the subtyping relation is strictly connected to the functionalities available through the definition in a class as export parameters. Then, subtyping is a more general relation than inheritance, and it is possible to specify when a class can be assumed as subtype of another class (Parisi Presicce and Pierantonio 1991).

The problem of automatic execution of specifications is also considered in TASSO. A particular model is proposed for the specifications, as sorts, signatures, and axioms which give meaning to the function symbols. The verification of properties of given objects is obtained by two different mechanisms: verification "a priori" and verification "a posteriori". In the case of "a priori" verification an editor oriented to the construction of the specification has been proposed. Executable code is then derived from the edited specifications (Antoy et al. 1990, Antoy 1990). For the "a posteriori" verification, a completeness checker for the rewriting rules associated with the specifications is available (Forcheri and Molfino 1990).

TASSO includes two main modules: TASSO-L and TASSO-D. TASSO-L is the language, based on an object-oriented paradigm, which allows the user to define objects, to instantiate abstract structures into actual computing domains, and to compute with available objects.

TASSO-D is the module for automated deduction. It can be used both by the user to check properties of given objects and by the system itself in order to generate and derive new properties of given objects from properties already known.

3 The language: TASSO-L

The TASSO language follows the object-oriented approach. According to the literature (e.g., Stroustrup 1987), this approach can be well characterized by the following equation

$$\text{Object-oriented} = \text{ADT} + \text{Inheritance} \ ,$$

and it appears particularly suitable to our purpose. In fact all the mathematical objects we like to consider, can be easily modelled by abstract data type (ADT). Furthermore, their implementation is encapsulated at a hidden level with respect to the user. The mathematical objects represented by ADT can be considered as generic structures in which data, methods and attributes are specified and localized.

The strong type checking mechanism, acting on all the defined objects, is the key to support the correctness of the types and subtypes used in all the different steps of a computation.

The available inheritance mechanism gives the possibility to specialize or extend an already defined ADT. In this way the typical hierarchy of mathematical objects is made correspondent to the tree of the inherited structures, which becomes embedded into the system.

Moreover, a correct cooperation between abstraction and inheritance allows also to obtain *parametric polymorphism*. Following this approach the common properties of similar data structures are defined at the highest possible level of abstraction and the methods to perform specific operations are dynamically defined only upon the appropriate operands.

According to the original motivation for TASSO, one of the most important characteristics to be considered is the necessity to maintain a high degree of correctness of the performed operations, also when dynamic data are defined: i.e., the flexibility of the language with respect to user's needs must be balanced by preventing incorrect or ambiguous types definitions. This fundamental objective can be obtained by a correct regulation of the instantiation mechanism of the dynamic data, together with a powerful type checker.

We have considered the most interesting, correct and flexible features of the existing object-oriented programming languages (Limongelli et al. 1990; Regio and Temperini 1990a, b). We have also been experimenting with some of them, namely Eiffel (Meyer 1992), Smalltalk (Ingalls 1978), C++ (Stroustrup 1986) and Loglan (Kreczmar et al. 1990). Recent results on this subject led to the definition of an Enhanced Strict Inheritance mechanism (Di Blasio and Temperini 1993, 1997) and to its implementation in TASSO-L (Di Blasio et al. 1997).

4 The automated deduction mechanisms: TASSO-D

The language TASSO-L incorporates automated deduction mechanisms: TASSO-D. Those mechanisms can be activated directly by the user when he has to accomplish some steps of deduction, and they have also to be automatically applied by the system itself as a tool to guarantee the correctness of the objects definitions.

The following are classes of properties treatable by the automated deduction mechanisms.

- those properties defined by the user during the specification of the objects: they concern the inheritance relations and can be tested during the compilation phase on the basis of the static tree of the prefixing structure and of the compatibility rules for the types;
- the properties of dynamically created objects: they must be verified after the creation of the objects in order to guarantee the correctness of the operations;
- the properties of an object generated by a computation: they must be verified on a user's request and can fall in two cases. In the *generation case* the user could ask for the properties of the object starting from known hypotheses. In the *causes-for-events case* the user can check the possible hypotheses under which known properties of the object have been obtained.

The TASSO-D module incorporates different deduction mechanisms. We have considered different automated deduction approaches. In the present version of TASSO object-oriented implementations of the *resolution* with different kinds of *backtracking* (Bonamico and Cioni 1988), of the *connection method* (Bibel 1987, Forcellese and Temperini 1990) and of a *sequent calculus* (Gallier 1986) are available.

These mechanisms offer a good degree of human orientation and in some cases their applicability is increased by the interactive use. The user has the knowledge of his application field, can modify the deduction path or stop it or, also, execute it under the control of a specific ad-hoc strategy. Therefore the interaction is useful in this perspective.

The connection method is based on the possibility of representing well-formed formulae (wffs) as matrices whose elements are (matrices of) atoms, and a validation of a formula is obtained by searching a path of connections between columns of the matrix, which represent the clauses of a disjunctive normal form (DNF) of the formula. Then, this method can be defined as an algorithmic mechanism operating upon a single data structure which directly corresponds to the given formula.

Some remarks can be made on the characteristics of this deduction mechanism.

- The efficiency and the transparency of the method derive from the possibility of working, during the entire deduction process, with the same unmodified matrix, created in the input step. In particular, it has to be stressed that any kind of preprocessing of the original formula (e.g., to transform it in a normal form) can be avoided.
- This mechanism can be considered flexible because it is extendible to higher order logics and also because the corresponding proof based on natural Gentzen calculus can be automatically obtained from a proof based on this connection method.
- The connection method, using a single data structure, in which the formula is completely encapsulated, is particularly suitable of an object-oriented implementation and in this way the uniformity with the entire system can be maintained.

The extension of the deduction capabilities to include the *abduction rule* in the *resolution* approach (as proposed by Kowalski 1979) is known in the literature. Similarly the extension from the verificative case to the causes for events case has been proposed for the connection method (Forcellese and Temperini 1990).

Furthermore the possibility of defining a single method to support different kinds of deductions, namely verificative, generative, and causes-for-events, has been considered. To this purpose a *sequent calculus* has been defined because this method results to be intrinsically suitable by being based on a set of rules easily defined and applied according to the different objective.

Starting from Gallier's proposal (Gallier 1986), we have defined a sequent calculus with a set of rules for both the decomposition of a given formula into elementary atoms, and for the composition of many sequents into a normal form. We started from the propositional case and we designed the kernel of a tool to be used also in the predicative case. We have used some strategies in order to augment the efficiency and to give some halting criteria to overcome the undecidibility of the general problem (Cioni et al. 1995).

5 Development and testing

The development of a complete TASSO demonstration is still in progress mainly on the integration of the different modules, described in the previous sections, and on the user's interface.

In fact, the current implementations of TASSO-L and TASSO-D have been widely tested as single moduli and their interaction and cooperation will be verified now on an appropriate set of examples of real applications.

The starting point of this experimentation has been proposed by D. Wang (1990) by stating a set of interesting problems together with some solution strategies. Those examples came from different areas of applied mathematics, such as irrational expressions simplification, stability of differential equations, linear algebra, limits, geometry reasonings, and number theory.

According to the original motivation for the TASSO project all these examples appear as impossible to be treated by the existing systems for mathematical problem solving and can be assumed as a good bed for the test of the system.

The development of TASSO has followed an incremental approach. Once a preliminary version of TASSO-L and of TASSO-D has been made running we started by defining some classes of elementary objects, such as integers, rationals, polynomials, matrices.

During this developing phase the automated deduction mechanism available in TASSO-D has been widely used to check properties of objects, to generate properties of objects under definition, to verify the correctness of the computing mechanism encapsulated into the objects.

Acknowledgements

This work has been partially supported by MURST under the projects "Calcolo Algebrico e Simbolico" and "Sistemi Intelligenti", and by CNR under the project "Sistemi Informatici e Calcolo Parallelo", grant no. 92.01604.69.

References

Antoy, S. (1990): Design strategies for rewrite rules. In: Kaplan, S., Okada, M. (eds.): Conditional and typed rewriting systems. Springer, Berlin Heidelberg New York Tokyo, pp. 333–341 (Lecture notes in computer science, vol. 516).

Antoy, S., Forcheri, P., Molfino, M. T. (1990): Specification-based code generation. In: Shriver, B. D. (ed.): Proceedings 23rd Hawaii International Conference on System Sciences, vol. II, software. IEEE Computer Society Press, Los Alamitos, pp. 165–173.

Bibel, W. (1987): Automated theorem proving. F. Vieweg und Sohn, Wiesbaden.

Bonamico, S., Cioni, G. (1988): Embedding flexible control strategies into object oriented languages. In: Mora, T. (ed.): Applied algebra, algebraic algorithms and error-correcting codes. Springer, Berlin Heidelberg New York Tokyo, pp. 454–457 (Lecture notes in computer science, vol. 357).

Buchberger, B., Collins, G. E., Loos, R. (eds.) (1983): Computer algebra, symbolic and algebraic computation, 2nd edn. Springer, Wien New York.

Caviness, B. (1986): Computer algebra: past and future. J. Symb. Comput. 2: 217–236.

Cioni, G., Colagrossi, A., Miola, A. (1995): A sequent calculus for symbolic computation systems. J. Symb. Comput. 19: 175–199.

Davenport, J. H., Siret, Y., Tournier, E. (1988): Computer algebra: systems and algorithms for algebraic computation. Academic Press, London.

Di Blasio, P., Temperini, M. (1993): Subtyping inheritance in languages for symbolic computation systems. In: Miola, A. (ed.): Design and implementation of symbolic computation systems. Springer, Berlin Heidelberg New York Tokyo, pp. 107–121 (Lecture notes in computer science, vol. 722).

Di Blasio, P., Temperini, M. (1997): On subtyping in languages for symbolic computation systems. In: Miola, A., Temperini, M. (eds.): Advances in the design of symbolic computation systems. Springer, Wien New York, pp. 164–178 (this volume).

Di Blasio, P., Temperini, M., Terlizzi, P. (1997): Enhanced strict inheritance in TASSO-L. In: Miola, A., Temperini, M. (eds.): Advances in the design of symbolic computation systems. Springer, Wien New York, pp. 179–195 (this volume).

Forcellese, G., Temperini, M. (1990): Towards a logic language: an object-oriented implementation of the connection method. In: Miola, A. (ed.): Design and implementation of symbolic computation systems. Springer, Berlin Heidelberg New York Tokyo, pp. 61–70 (Lecture notes in computer science, vol. 429).

Forcheri, P., Molfino, M. T. (1990): Educational software suitable for learning programming methodologies. In: Onate, E., Suarez, B., Owen, D. R. J., Schrefler, B., Kroplin, B., Kleiber, M. (eds.): Proceedings Conference on Computer Aided Training in Science and Technology, CATS '90. Centro International de Metodos Numericos en Ingeneria, Barcelona, pp. 161–164.

Gallier, J. H. (1986): Logic for computer science. Harper and Row, New York.

Ingalls, D. (1978): The smalltalk 76 programming system design and implementation. In: Proceedings Symposium on Principles of Programming Languages. Association for Computing Machinery, New York, pp. 9–16.

Kowalski, R. (1979): Logic for problem solving. North-Holland, Amsterdam.

Kreczmar, A., Salwicki, A., Warpechowski, M. (1990): LOGLAN '88 – report on the programming language. Springer, Berlin Heidelberg New York Tokyo (Lecture notes in computer science, vol. 414).

Limongelli, C., Miola, A. (1990): Abstract specification of numeric and algebraic computation methods. In: Balagurusamy, E., Sushila, B. (eds.): Computer systems and applications, recent trends. Tata McGraw-Hill, New Delhi, pp. 27–35.

Limongelli, C., Temperini, M. (1992): Abstract specification of structures and methods in symbolic mathematical computation. Theor. Comput. Sci. 104: 89–107.

Limongelli, C., Mele, M. B., Regio, M., Temperini, M. (1990): Abstract specification of mathematical structures and methods. In: Miola, A. (ed.): Design and implementation of symbolic computation systems. Springer, Berlin Heidelberg New York Tokyo, pp. 61–70 (Lecture notes in computer science, vol. 429).

Meyer, B. (1992): Eiffel: the language, 2nd edn. Prentice Hall, Englewood Cliffs.

Miola, A. (ed.) (1990a): Computing tools for scientific problems solving. Academic Press, London.
Miola, A. (ed.) (1990b): Design and implementation of symbolic computation systems. Springer, Berlin Heidelberg New York Tokyo (Lecture notes in computer science, vol. 429).
Miola, A. (ed.) (1993a): Design and implementation of symbolic computation systems. Springer, Berlin Heidelberg New York Tokyo (Lecture notes in computer science, vol. 722).
Miola, A. (1993b): Symbolic computation systems. In: Kent, A., Williams, J. G. (eds.): Encyclopedia of computer science and technology. Marcel Dekker, New York, pp. 367–380.
Miola, A., Mora, T. (1988): Constructive lifting in graded structures: a unified view of Buchberger and Hensel methods. J. Symb. Comput. 6: 305–322.
Parisi Presicce, F., Pierantonio, A. (1991): An algebraic view of inheritance and subtyping in object-oriented programming. In: van Lamsweerde, A., Fugetta, A. (eds.): ESEC '91. Springer, Berlin Heidelberg New York Tokyo, pp. 364–379 (Lecture notes in computer science, vol. 550).
Pavelle, R. (1985): Applications of computer algebra. Kluwer, Boston.
Regio, M., Temperini, M. (1990a): Implementation and manipulation of formal objects: an object-oriented view. In: Proceedings ACM Symposium on Personal and Small Computers, Arlington, VA USA, March 28–30, 1990. Association for Computing Machinery, New York.
Regio, M., Temperini, M. (1990b): Object-oriented methodology for the specification and the treatment of mathematical objects. In: Balagurusamy, E., Sushila, B. (eds.): Computer systems and applications, recent trends. Tata McGraw-Hill, New Delhi, pp. 94–103.
Stroustrup, B. (1986): An overview of C++. SIGPLAN Not. 21/11: 7–18.
Stroustrup, B. (1987): What is object-oriented programming. In: Bèzivin, J., Hullot, J.-M., Cointe, P., Liebermann, H. (eds.): Proceedings European Conference on Object-Oriented Programming. Paris, France, June 1987, pp. 51–60.
Wang, D. (1990): Some examples for testing an integrated system. Tech. Rep. 4/18, P. F. Sistemi Informatici e Calcolo Parallelo, Consiglio Nazionale delle Ricerche, Rome.

Mathematical objects

The uniform representation of mathematical objects by truncated power series

C. Limongelli and M. Temperini

1 Introduction

Systems for symbolic mathematics are based on the availability of powerful methods and techniques, which have been developed for numeric computation, symbolic and algebraic computation and automated deduction. But those different computing paradigms really work independently in such systems. Each of them represents an individual computing environment, while they are not integrated to support a uniform environment for computation. The problem of the integration of numeric and symbolic computation is still open (Caviness 1986, Limongelli and Miola 1990, Mascari and Miola 1986).

In this paper, a possible solution to this problem is considered, taking into account the well-known need for abstraction in the definition of mathematical data structures. Mainly we focus on the possibility of using a uniform representation of those structures and of extending the use of algebraic algorithms to numerical settings.

With this aim in mind, we will present a classification of formal structures, defined in an algebraic context. In such a classification a fundamental role is played by the methods defined on algebraic structures: the algorithms for the manipulation of abstract structures can be included in the definition of the structures themselves. They act as polymorphic operators and can be applied on every structure which is compatible with the structure in which the operator is defined, and which is defined at a lower level of abstraction.

This approach is based on abstraction techniques and on the object-oriented paradigm. Hence methods are considered from a two-fold viewpoint: first they are interpreted as characterizing attributes of the entity on which they are defined; second they can provide the algorithms that may be not characterizing attributes, but enrich the definition of the structure in which they are given.

Related to the former case, e.g., we can cite the operators $+$ and $*$, defined in a ring structure; these methods are necessary for the definition of this structure. Related to the latter case, we can consider, e.g., the Hensel lifting algorithm (see Sect. 2); this algorithm is joined to the definition of the most abstract structure on which it acts. In such a way this algorithm can be applied on any instance of the abstract structure and on all descendant structures.

Moreover, this classification allows us to identify the algebraic structure from which it is possible to specialize numbers and functions. We define $\mathbb{D}[[x]]$ as the truncated power series with coefficients defined over a domain $\mathbb{D}$ and $\mathbb{D}[x]$ the univariate polynomials defined over $\mathbb{D}$. It is well known that every analytical function can be represented as a succession of truncated power series. Moreover we know that $\mathbb{D}[[x]]$ is isomorphic to $\mathbb{D}[x]$.

The proposed representation for functions and numbers is based on truncated power series. It allows, through the p-adic arithmetic tool (Colagrossi et al. 1997), for the integration of the numeric and symbolic capabilities of algorithms defined at high level of abstraction. In this work we describe how a uniform representation for functions and numbers, based on the p-adic construction, can be used in order to obtain a homogeneous approach to symbolic and numeric computation.

The well known Hensel algorithm is extended to cover both symbolic and numeric operations: polynomial factorization, the n-th root of an analytical function, root finding of polynomial equations, and p-adic expansion of an algebraic number are examples of operations that can be performed by this extension. The extended algorithm works on a common p-adic representation of algebraic numbers and functions. In the numeric case the computations are performed by the p-adic arithmetic. Actually, numeric computations by the extended Hensel algorithm provide the exact representation of the resulting algebraic numbers, by means of their p-adic expansion.

2 Classification of mathematical objects

The specification of an algebraic structure is defined through the notion of signature. A signature Σ consists of a triple $\langle S, O, P\rangle$: S is a disjoint union of sets (sorts) on which the corresponding algebra is defined; O is a proper set of operators (function symbols) such that each operator is associated to a mapping type $s_1 \ldots s_k \to s$ where $s_1 \ldots s_k$ and s belong to S; P is a set of properties defining the relationships among operators, expressed through a logic formalism, e.g., by formulae of first-order logic (Wirsing and Broy 1989), or expressions of algorithmic logic (Limongelli et al. 1990). The choice of an axiomatic specification mechanism is taken in order to satisfy the design requirements of a symbolic computation system (Limongelli et al. 1992), as specified in Sect. 1.

In the following a classification of the algebraic structures is proposed, defining three planes for their static definition. This subdivision will point out the different specification requirements of structures lying on different planes, leading to a higher level of correctness in their treatment. In the sequel the term symbolic computation will be used for any computation involving the manipulation of properties of a structure, without worrying about the abstraction level of its definition. The term numeric computation will refer to computation that acts over structures that have completely specified sorts. Moreover, it will be shown how object-oriented programming (OOP) methodology naturally fits into the specification scheme outlined above.

2.1 Abstract structures

The definition of algebraic structures is carried out by a given scheme supporting the declaration of sorts, operators, and their related axioms. The first step towards the definition of such a scheme should include mechanisms for the classification of the algebraic structures that are found at the highest level of abstraction as stated in classical algebra. In this way a hierarchy of structures (see Fig. 1) is established. The elements of this hierarchy will be named *abstract structures*. The hierarchy specializes itself by adding new properties and operations, starting from a very general structure (e.g., semigroup). From now on the single arrows in the figures represent inheritance relations: the direction of an arrow is intended as starting from the "inherited" class *(parent)* towards the "inheriting" one *(successor)*.

At a general specification level, as in classical algebra, sorts are not specified, while only operations and axioms, related to the presentation of an algebraic structure, are defined. Abstract structures cannot be employed for numeric computations: only symbolic computations can be carried out on them. In fact, numeric computations are performed only when the sorts of the involved algebraic structures are completely specified. In Table 1 the specification of the algebraic abstract structures of Fig. 1 is shown. In this specification the possibility of obtaining the definition of a structure taking sorts, operators, and properties from another definition, possibly by redefining some denotations, is exploited.

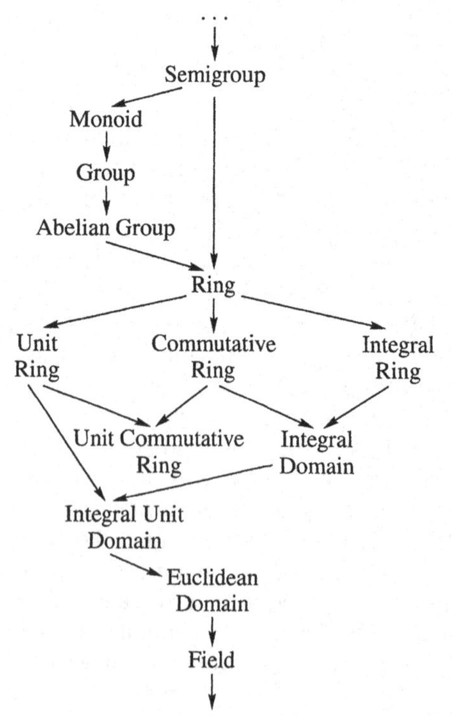

Fig. 1. Classification of abstract algebraic structures

Table 1. Abstract structure specification

Semigroup	
sorts	\$
operations	op: $\$ \times \$ \to \$$
properties	$\forall a, b, c \in \$,\ (a \operatorname{op} b) \operatorname{op} c = a \operatorname{op}(b \operatorname{op} c)$
Monoid	
from	*Semigroup*
operations	1: $\to \$$
properties	$a \in \$,\ 1 \operatorname{op} a = a \operatorname{op} 1 = a$
Group	
from	*Monoid*
properties	$\forall a \in \$,\ \exists a^{-1} \in \$\colon\ a \operatorname{op} a^{-1} = a^{-1} \operatorname{op} a = 1$
Abelian Group	
from	*Group*
properties	$a, b \in \$,\ a \operatorname{op} b = b \operatorname{op} a$
Ring	
from	*Abelian Group* redefining op as $+$, 1 as 0, a^{-1} as $-a$
from	*Semigroup* redefining op as $*$
properties	$\forall a, b, c \in \$,\ a * (b + c) = a * b + a * c$
	$\forall a, b, c \in \$,\ (a + b) * c = a * c + b * c$
Unit Ring	
from	*Ring*
operations	1: $\to \$$
properties	$\forall a \in \$,\ a * 1 = 1 * a = a$
Commutative Ring	
from	*Ring*
properties	$\forall a, b \in \$,\ a * b = b * a$
Comm. Unit Ring	
from	*Unit Ring*
from	*Commutative Ring*
Integral Ring	
from	*Ring*
properties	$\forall a, b \in \$,\ a * b = b * a = 0 \iff a = 0 \vee b = 0$
Integral Domain	
from	*Commutative Unit Ring*
from	*Integral Ring*
Euclidean Domain	
from	*Integral Ring*
operations	div: $\$ \times \$ \to \$$, mod: $\$ \times \$ \to \$$
properties	$\forall a, b, c \in \$,\ \exists c, r \in \$\colon\ (a = c * b + r) \Rightarrow$
	$(a \operatorname{div} b = c) \wedge (a \operatorname{mod} b = r)$
Field	
from	*Integral Domain*
properties	$\forall a \in \$,\ a \neq 0,\ \exists a^{-1}\colon\ a * a^{-1} = a^{-1} * a = 1$

Limongelli and Temperini (1992) showed how this characteristic, which is classical in algebra, will match directly with inheritance in OOP.

It should be noted how in this specification the use of the inheritance mechanism by means of "from" construction makes it possible to obtain shorter but equally expressive definitions.

2.2 Parametric structures

In order to actually use the specified abstract structures, the need arises for more specialized properties and operations. A first step towards this objective is the specification of an intermediate level of specialization, where there are essentially two requirements:

- the specialization of the sorts which is accomplished by the specification of sorts containing parameters, i.e., the complete specification of sorts depends on the specification of other sorts;
- the complete semantic definition of the operations, i.e., the methods which are characteristic of these structures.

These structures will be named *parametric structures*. Matrices, Polynomials and Formal Power Series are examples of parametric structures. The correspondence with the related abstract structures is shown in Fig. 2. Here, for the sake of clarity, a distinction among abstract and parametric structures is outlined by placing these structures on different planes: an *abstract plane* and a *parametric plane*, respectively.

Let us consider the case of the matrix structure. A general square matrix of

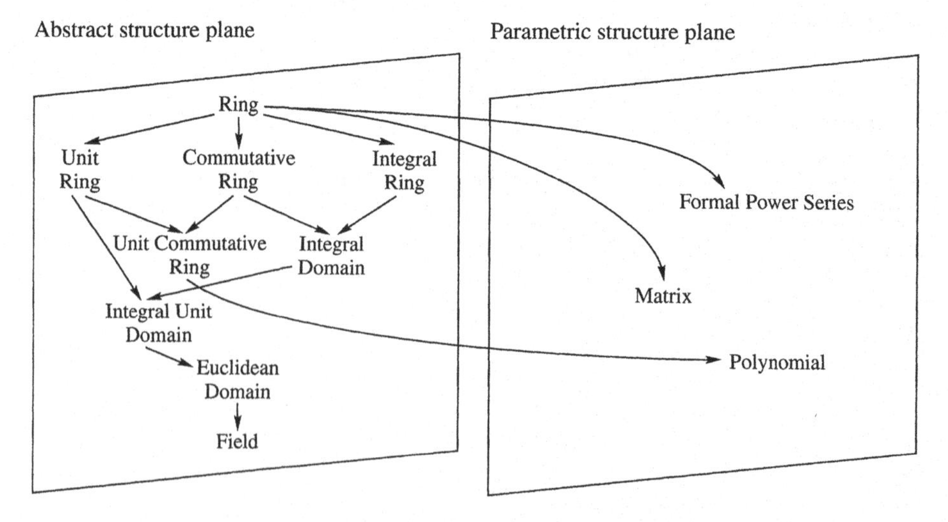

Fig. 2. Example of parametric structure

order n (n is a natural number) is an element of the set

$$\mathrm{MAT} = \{\{a_{i,j}\}_{i,j \in I_n},\ a_{i,j} \in \$\}\ ,$$

where

$$I_n = i \in \mathrm{NAT}\colon\ 1 \leq i \leq n$$

and NAT is the sort of natural numbers.

Definition 1 (Matrix structure specification).
Matrix

sorts NAT, I_n, \$, *Ring*

operations from *Ring* on MAT redefining $+$ as $+_{\mathrm{MAT}}$ and $*$ as $*_{\mathrm{MAT}}$
put: MAT × NAT × NAT × \$ → MAT
get: MAT × NAT × NAT → \$

properties $\forall i, j \in I_n,\ \mathrm{get}(0_{\mathrm{MAT}}, i, j) = 0_\$$;
$\forall M \in \mathrm{MAT},\ \forall i, j \in I_n,\ \forall e \in \$,\ \mathrm{get}(\mathrm{put}(M, i, j, e), i, j) = e$;
$\forall M_1, M_2 \in \mathrm{MAT}$,
$M_1 +_{\mathrm{MAT}} M_2 = \{\mathrm{get}(M_1, i, j) +_\$ \mathrm{get}(M_2, i, j)\}_{i,j,\in I_n}$;
$\forall M_1, M_2 \in \mathrm{MAT}$,
$M_1 *_{\mathrm{MAT}} M_2 = \{\mathrm{get}(M_1, i, 1) *_\$ \mathrm{get}(M_2, 1, j) +_\$ \ldots +_\$$
$+_\$ \mathrm{get}(M_1, i, n) *_\$ \mathrm{get}(M_2, n, j)\}_{i,j,\in I_n}$;

The use of the "from" construction with the specification "on MAT" means that the operations ($+$ and $*$), inherited from the Ring structure, are applied to the parametric sort MAT: they will be denoted respectively as $+_{\mathrm{MAT}}$ and $*_{\mathrm{MAT}}$. In Definition 1, only a partial specification of matrix coefficients is given by the "parameter sort" \$ (so the coefficients are elements of a Ring structure in which 0_{MAT} is the null element for $+_{\mathrm{MAT}}$; $0_\$$ represents the null element of \$).

2.3 Ground structures

When sorts are completely specified, the resulting structure allows a direct variable instantiation (in this case the structure will be named *ground*). Therefore it can be used both for symbolic and numeric computations. Moreover it can be noted that new computational methods and properties can be added into the definition of a ground structure, in order to reach a higher level of efficiency, both for computation and deduction (see Fig. 3).

The characteristics of the structures defined above are summarized as follows:

- *abstract structures:* classical algebraic structures described by inherited properties. Sorts are not specified, only symbolic computations can be performed;
- *parametric structures:* e.g., matrix structure, enrich the definition of abstract structures by partial (parameterized) sorts, and by additional operations and properties;

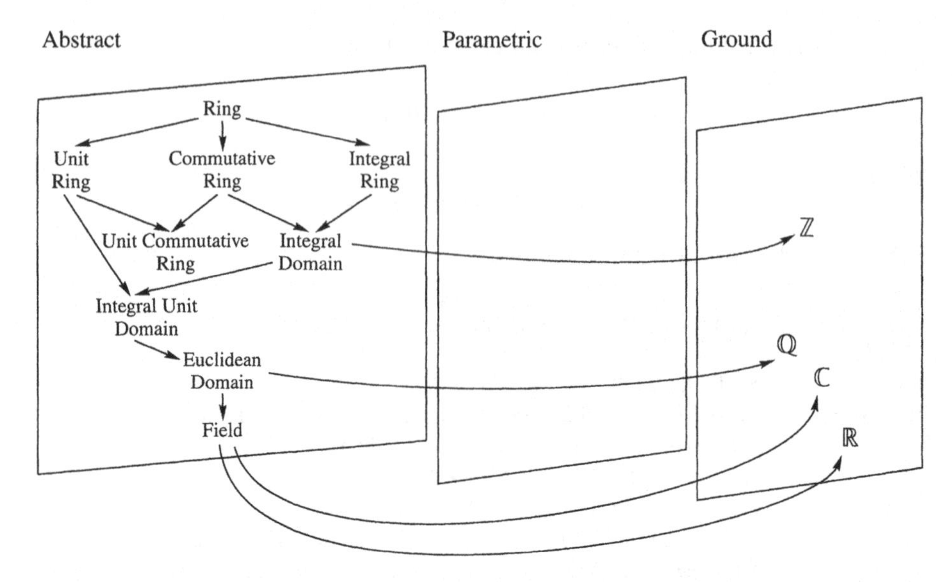

Fig. 3. Plane of ground structures

- *ground structures:* e.g., integers, completely specified; both symbolic and numeric computations are allowed.

The *definition space* is defined by the three previously described planes. Here two different levels of hierarchy can be imagined: a first level is limited to the plane of abstract structures, and defines their hierarchy; a second interplanar level acts respectively between the abstract and parametric structure planes and between abstract and ground structure planes. The *execution space* takes place out of the definition planes; that is the space where the structures are instantiated into algebraic objects. This space represents the connection between a single object and its generating structure. In Fig. 4 the bold arrows stand for object instantiation.

For example an integer number is joined with its ground structure $\mathbb{Z}$, a matrix of integers is joined both with a Matrix parametric structure and $\mathbb{Z}$ ground structure. A matrix of integer polynomials is joined with the structure $\mathbb{M}(\mathbb{P}(\mathbb{Z}))$.

It must be noted that an abstract structure is never instantiated. The chain of inheritance acts only on the abstract structure plane. On the other hand, the elements that belong respectively to the parametric structure plane or to the ground structure plane are not themselves related. Their correlations are established by transversal inheritance from the abstract plane to the parametric and ground planes.

3 Method abstractions

In the previous section it has been seen that algebraic structures present some method definitions. In general, these methods specify the semantics of fundamental operations in the structure (e.g., the matrix addition method in the Matrix

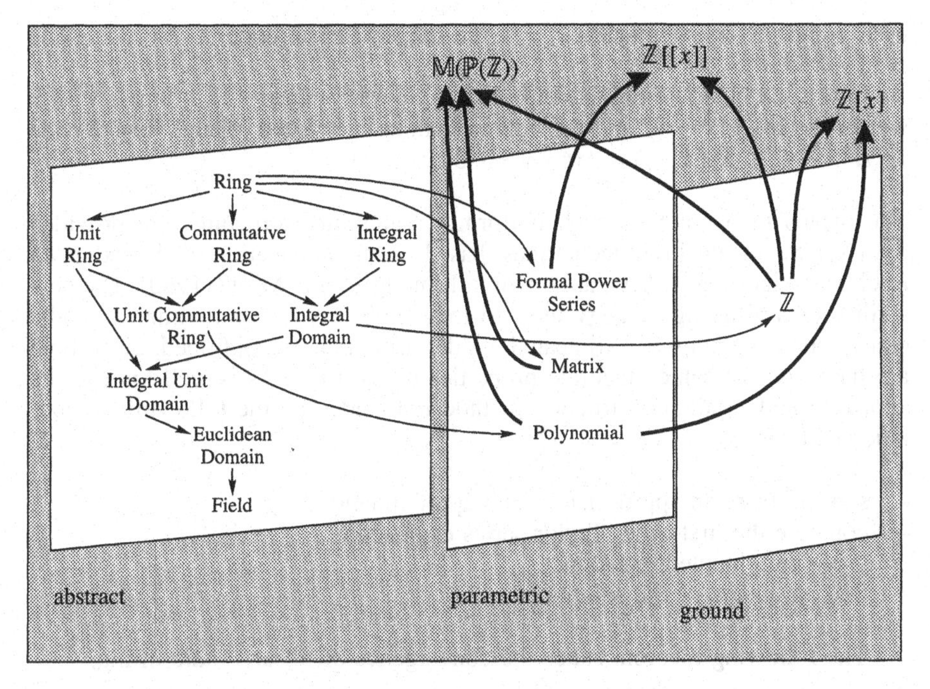

Fig. 4. The execution space

parametric structure, or the "div" and "mod" methods defined in the Euclidean Domain abstract structure).

Now it will be shown how the enrichment of the definition of algebraic structures is possible, thus giving rise to abstract structures in which computing methods become available components. In fact, as far as method abstraction is concerned, the inheritance allows us to use the code provided for a higher structure in all its subdomains, without any redefinition, as Limongelli et al. (1990) and Limongelli and Temperini (1992) show.

The approach to numeric computation through truncated p-adic arithmetic (Hensel 1908) has been widely analyzed (Gregory and Krishnamurthy 1984, Colagrossi et al. 1997). As a matter of fact, the truncated p-adic representation can be viewed as the truncation of the p-adic series representing the rational number. The degree r, chosen for the approximation, depends on the characteristics of the problem. The same approach can be used to treat problems in abstract structures. In particular, it is possible to exploit the uniformity between the truncated p-adic expansion of a rational number (Hensel code) and the p-adic polynomial expansion by power series. The theory and construction of the Hensel code for arithmetic of a rational function over a finite field is analogous to the truncated p-adic arithmetic. So a rational polynomial function can be represented in the form of the Hensel code as Example 1 shows.

Example 1. Given the univariate polynomial $P(x) = x^3+3x^2-2x+1$ we fix the base x and the approximation $r = 4$, the related Hensel code is: $H(x, 4, p(x)) = (1\ -2\ 3\ 1, 0)$.

Given the bivariate polynomial $P_1(x, y) = yx^2 - 2yx + 2y + y^2x - y^2 - 3$ we fix the base $(x - 1)$ and the approximation $r = 4$, the related Hensel code is: $H(x - 1, 4, P_1(x, y)) = (y^2\ y\ y\ -3, 0)$.

The means to operate on such uniformly represented structures are provided by a variety of algebraic techniques, based on the application of the extended Euclidean algorithm (EEA), as shown in Limongelli and Miola (1990). The possibility of a unifying strategy used to treat problems by approximated p-adic construction methods and truncated p-adic arithmetic is confirmed, since both are based on the same algebraic properties of the EEA. As is well known, the approximated p-adic construction method is founded on the following computational steps:

- starting from an appropriate initial approximation;
- compute the first order Taylor series expansion;
- solve the obtained equation;
- find an update of the solution.

The following theorem states the convergence of linear p-adic lifting.

Theorem 1 (Abstract linear lifting). Let I be a finitely generated ideal in R, $f_1, \ldots, f_n \in R[x_1, \ldots, x_r]$, $r \geq 1$, $a_1, \ldots, a_r \in R$ with

$$f_i(a_1, \ldots, a_r) \equiv 0 \bmod I, \quad \text{with } i = 1, \ldots, n .$$

Moreover let

$$U = (u_{i,j}), \ \ i = 1, \ldots, n, \ \ j = 1, \ldots, r, \quad \text{with } u_{i,j} = \frac{\partial f_i}{\partial x_j}(a_1, \ldots, a_r) \in R$$

(U is the Jacobian matrix of $f_1, \ldots, f_n$, evaluated at $a_1, \ldots, a_r$). Assume that U is invertible mod I.

Then, for any positive integer t, there exist $a_1^{(t)}, \ldots, a_r^{(t)} \in R$, such that

$$f_i(a_1^{(t)}, \ldots, a_r^{(t)}) \equiv 0 \bmod I^t, \quad i = 1, \ldots, n$$

and

$$a_j^{(t)} \equiv a_j \bmod I, \quad j = 1, \ldots, r .$$

Proof. The proof is given by induction on t. See Lauer (1983). □

The Hensel lemma is a specialization of that theorem. The lemma is specialized by assuming $n = 1$, $r = 2$, $R = \mathbb{Z}[x]$ and $p = I$, p being a prime number. Thus, assume $A_1, A_2, C \in \mathbb{Z}[x]$. Rename A_1 and A_2 as G and H, respectively, and consider $C = F$. Also suppose that

$$G \cdot H \equiv F \bmod p ,$$

and that G and F are relatively prime mod p. Then, for any positive integer t there exist $A_1^{(t)}, A_2^{(t)} \in \mathbb{Z}_{p^t}[x]$, such that

$$G^{(t)} \equiv G \bmod p \quad \text{and} \quad H^{(t)} \equiv H \bmod p .$$

Moreover, since $n = 1$, we can rename f_1, Φ. Thus we consider

$$\Phi(G, H) = G \cdot H \equiv 0 \bmod p .$$

In such case the partial derivatives are

$$u_{1,1} = \frac{\partial \Phi}{\partial G}(G, H) = H ,$$

$$u_{1,2} = \frac{\partial \Phi}{\partial H}(G, H) = G$$

and the matrix U (1×2) is equal to (H, G) and is invertible mod p, if and only if G and H are relatively prime mod p.

If $G^{(t)}, H^{(t)}$ satisfies

$$G^{(t)} H^{(t)} - F \equiv 0 \bmod p^t ,$$

and if we consider

$$G^{(t+1)} = G^{(t)} + Ap^t, \quad H^{(t+1)} = H^{(t)} + Bp^t ,$$

by solving the two equations in the indeterminates A and B, we obtain:

$$A \cdot H + B \cdot G \equiv 0 \bmod p .$$

The specialization described above can be reformulated by the following lemma. This lemma results from Theorem 1 renaming t as k, $G^{(t)}$ and $H^{(t)}$ as G_k and H_k respectively.

Lemma 1 (Hensel lemma). Given a prime number $p \in \mathbb{Z}$, and given $F(x) \in \mathbb{Z}[x]$, if $G_1(x)$ and $H_1(x)$ are two polynomials relatively prime over $\mathbb{Z}_p[x]$, such that

$$F(x) \equiv G_1(x) \cdot H_1(x) \bmod p ,$$

then, for any integer $k \geq 1$ there exist two polynomials $G_k(x), H_k(x) \in \mathbb{Z}_{p^k}[x]$ such that

$$F(x) \equiv G_k(x) \cdot H_k(x) \bmod p^k$$

where

$$G_k \equiv G_1(x) \bmod p, \qquad H_k \equiv H_1(x) \bmod p \ .$$

Proof. See Yun (1974). □

This lemma ensures that, starting from an appropriate initial approximation, we can obtain higher order approximations.

Following Newton's idea (Lipson 1976), the Hensel method gives an iterative mechanism to solve equations of the following type: $\Phi(G, H) = 0$ where

$$\Phi\colon \mathbb{D}[x] \times \mathbb{D}[x] \to \mathbb{D}[x]$$

with $\mathbb{D}$ being an abstract Euclidean domain.

The generality of this method makes it an apt tool to deal with approximated representations in a uniform way and in the same algebraic context. The Hensel algorithm can perform different computational methods according to different actual specializations of the parameters G and H (i.e., different forms of the equation and of the initial approximations). The following scheme provides some examples:

i. factorization:	$\Phi(G, H) = F - GH = 0$;
ii. n-th root of a function G:	$\Phi(G) = F - G^n = 0$;
iii. Legendre polynomial:	$\Phi(G, t) = (1 - 2x + t^2)G^2 - 1 = 0$;
iv. Newton's method:	$\Phi(x) = F = 0$.

Starting from the initial approximations, $G_1(x)$ and $H_1(x)$ such that $\Phi(G_1, H_1) \equiv 0 \bmod I$, at each step k the approximate solution G_k, H_k can be obtained, such that $\Phi(G_k, H_k) \equiv 0 \bmod I^k$, where I is an ideal in $\mathbb{D}[x]$. Then the bivariate Taylor series expansion of Φ at the point (G_k, H_k), is computed as:

$$\Phi(G, H) = \Phi(G_k, H_k) + (G - G_k)\frac{\partial\Phi(G_k, H_k)}{\partial G} + (H - H_k)\frac{\partial\Phi(G_k, H_k)}{\partial H} \ .$$

Dividing by p^k and assuming

$$C = \Phi(G_k, H_k), \quad A_k = -\frac{\partial\Phi(G_k, H_k)}{\partial G}, \quad B_k = -\frac{\partial\Phi(G_k, H_k)}{\partial H},$$
$$\frac{G - G_k}{p^k} = \Delta G_k, \quad \frac{H - H_k}{p^k} = \Delta H_k \ ,$$

we solve the following equation

$$A_k \Delta G_k + B_k \Delta H_k \equiv \frac{C}{p^k} \bmod p \tag{1}$$

in order to find the $(k+1)$-th updates

$$G_{k+1} = G_k + p^k B_k, \qquad H_{k+1} = H_k + p^k A_k \ .$$

Here G_i (respectively H_i) stands for the sum of the first i terms in the p-adic series expansion of G (respectively H).

A more detailed description of the Hensel algorithm is reported in Limongelli and Temperini (1992). At each step we rebuild the p-adic coefficients related to series expansion of the solution. We must note that the hypothesis of the relative primality of G_1 and H_1 in the previous lemma is necessary only to solve the above Diophantine equation (1). When F has a specialized form, like in (ii), (iii), and (iv), the equation to be solved is just a modular univariate equation. In these cases the hypothesis of relative primality is no longer needed. A unifying view of different data structures and algebraic methods comes out by their treatment through a generalized p-adic representation. Figure 5 shows a set of application areas of the Hensel algorithm, distinguishing the numeric from the symbolic cases.

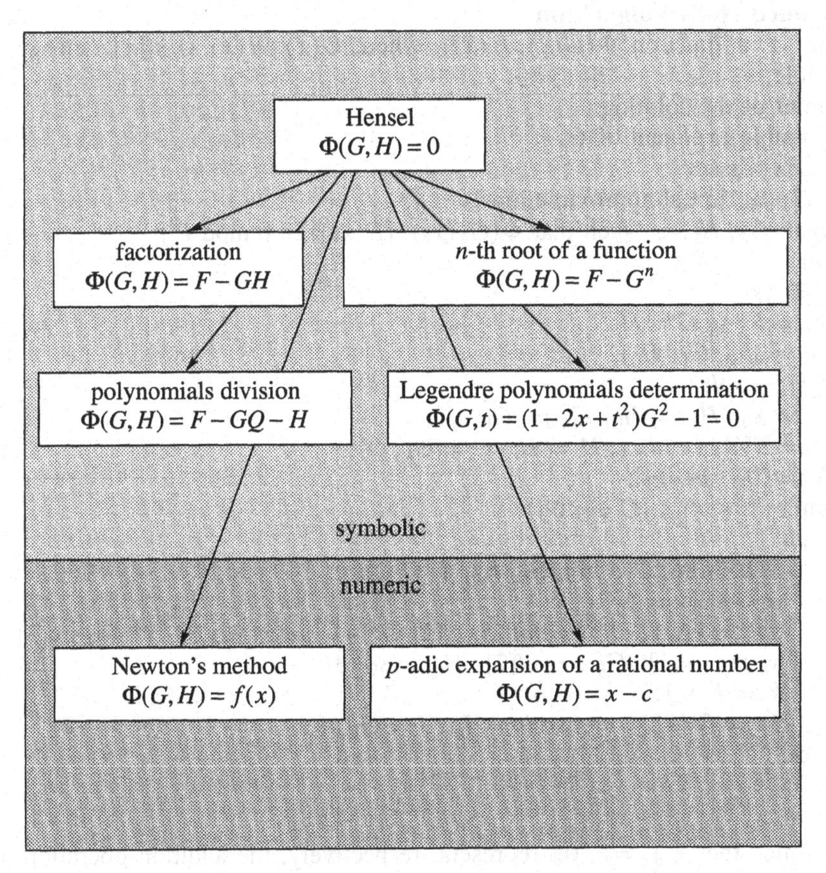

Fig. 5. Possible specializations of the Hensel method

3.1 Placing the Hensel method at the right specification level

Once the treatment of numeric data structures is founded on the p-adic representation and arithmetic, it is possible to devise a hierarchy of data structures, which represents the computational domain of symbolic (and numeric) computations. An approach, based on abstraction, to the classification of algebraic structures has been described in Limongelli and Temperini (1992). Abstract structures (like ring or field algebraic structures), parametric structures (like matrix of ring elements, or polynomials over ring coefficients) and ground structures (like integer or real) can be distinguished by the different completion of their algebraic definition. Among all these structures, the mechanism of strict inheritance plus redefinition allows to derive each one by another. Following this approach, we can locate in the appropriate planes the algebraic methods and structures, which have been cited previously as main devices in the unified view of symbolic and numeric computations. The first structure is the truncated power series. It is straightforwardly a parametric structure whose coefficients are defined over a Euclidean domain.

Extended Hensel algorithm
Input: a: a function $\Phi(G(x), H(x))$, where $G(x), H(x) \in \mathbb{D}[x]$, and x is a variable;
r: order of the solution;
n: possible exponent of G;
$I \in \mathbb{D}[x]$: base;
G_0, H_0, x_0: initial approximations;
Output: G_k, H_k, x_k, such that $\Phi(G_k(x_k), H_k(x_k)) \equiv 0 \bmod I^k$;
begin
  $k := 1$;
  $\Phi := F -_{\mathbb{D}} G^n \cdot H$;
  $sol := [x_0, G_0, H_0]$;
  $a[1] := sol$;
  $x_k := x_0$; $H_k := H_0$; $G_k := G_0$;
  $C := EVAL(x = x_k, H = H_k, G = G_k, \Phi)$;
  $\Delta := [0_{\mathbb{D}}, 0_{\mathbb{D}}, 0_{\mathbb{D}}]$;
  while $(k \leq r) \wedge (C \neq_{\mathbb{D}} 0_{\mathbb{D}})$
    do
      $SOLVE(F, G, x, x_0, G_0, H_0, \Phi, C, I)$;
      $a[k+1] := \Delta$;
      $sol := [\Delta[1] \cdot I^k +_{\mathbb{D}} sol[1], \Delta[2] \cdot I^k +_{\mathbb{D}} sol[2], [\Delta[3] \cdot I^k +_{\mathbb{D}} sol[3]]$;
      $x_k := sol[1]$; $G_k := sol[2]$; $H_k := sol[3]$;
      $k := k + 1$;
      $C := EVAL(x = x_k, H = H_k, G = G_k, \Phi)$;
    od
end;

Notice that $+_{\mathbb{D}}$, $=_{\mathbb{D}}$, $0_{\mathbb{D}}$ represent, respectively, the addition operation, the equality relation, and the zero element in the domain $\mathbb{D}$. The function call

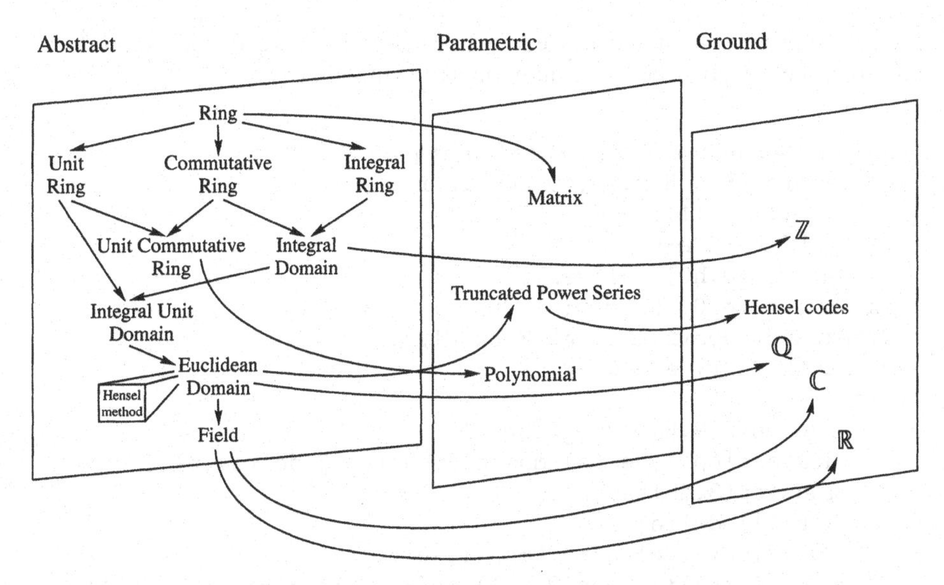

Fig. 6. Specification levels

EVAL($x = xk, H = Hk, G = Gk, F$) evaluates Φ in the point (G_k, H_k, x_k). The function *SOLVE* computes the expansion in Taylor series of a function F, and the related zeros. In particular, these zeros are computed w.r.t. either the symbolic variables (G, H) or the numeric variable (x) depending on the configuration of the given input parameters. Hence the algorithm is able to compute either symbolic or numeric solutions of Φ.

From the previous description we can see that the Hensel method takes polynomials as input and returns either a truncated power series or a Hensel code. It applies the EEA (in *SOLVE*, in order to solve either the Diophantine equation or the modular equation). It should be located where all the properties of a Euclidean domain are available (see Fig. 6). Hence the Euclidean domain is the structure of highest level of abstraction, where the Hensel method can be defined. Let us note that, in this case, the algorithm is not a characterizing attribute of the abstract structure of Euclidean domain; in fact, it is added to the definition, just enriching it.

The p-adic arithmetic is represented by the Hensel code which is conveniently located in the plane of ground structures. Hensel codes are particular instances of truncated power series whose coefficients are defined over $\mathbb{Z}_p$ (p being a prime number).

4 The integration of symbolic and numeric computation: a case study

In this section we propose a first implementation of the "integrated" extended Hensel algorithm. The programming language included in the system Maple (Char et al. 1986) has been used. The procedure `hgen` implements the plain

Hensel algorithm. In it the procedure hsolve, which we do not show here, performs the selection between numeric and symbolic case.

Implementation of extended Hensel algorithm

```
hgen := proc(F,G,H,x,x0,g0,h0,n,p,r);
  a:=array(1..r+1);
  k:=1; FI:=F-G^n*H;
  sol:= [x0,g0,h0]; a1:=sol;
  xk:=x0; gk:=g0; hk:=h0;
  C:=expand(subs(x=xk,G=gk,H=hk,FI));
  while (k <= r) and C<>0
    do
      hsolve(F,G,H,x,x0,g0,h0,FI,C,p);
      delta:=[op(1,delta) mod p, mods(op(2,delta),p);
      mods(op(3,delta),p)];
      a[k+1]:=delta;
      sol:=[op(1,delta)*p^k+op(1,sol);
      op(2,delta)*p^k+op(2,sol),op(3,delta)*p^k+op(3,sol)];
      xk:=op(1,sol); gh:=op(2,sol); hk:=op(3,sol);
      k:=k+1;
      C:=expand(subs(x=xk,G=gk,H=hk,FI));
    od
  for i to k do print(a[i]); od;
  if C=0 then print('exact result'); print(sol);
  else print('sum of first',k,'coefficients:',sol);
  fi;
end;
```

Actually, as in every other existing system for symbolic computation, the specification of data structure is not allowed as described in Sect. 3. So we had to design special structures in order to simulate the organization of abstract parametric and ground planes. In the rest of this section we will provide examples. Firstly we show the specialization of the Hensel algorithm in the symbolic case.

Let us show some examples of applications.

Example 2 (Factorization). Given the following univariate polynomial;

$$F(x) = x^5 + 12x^4 - 22x^3 - 163x^2 + 309x - 119$$

we want to compute its factorization in $\mathbb{Z}_5[x]$. Following the steps previously described we obtain:

1. starting from the initial approximations

$$G_1(x) = x^3 + 2, \;\; H_1(x) = x^2 + 2x - 2\;;$$

2. we compute the first-order bivariate Taylor series expansion

$$\frac{F - G_1 \cdot H_1}{p} - \left(\frac{G - G_1}{p}\right) \cdot H_1 - \left(\frac{H - H_1}{p}\right) \cdot G_1 \equiv 0 \bmod p \; ;$$

3. then we solve the Diophantine equation

$$\frac{F - G_1 \cdot H_1}{5} \equiv B_1(x^2 + 2x - 2) + A_1(2x - 1) \bmod 5$$

   finding the values

$$B_1 = 2x - 2; \;\; A_1 = 2x - 1; \;\; A_1, B_1 \in \mathbb{Z}_5[x] \; ;$$

4. then we update the initial solutions:

$$G_2 = (x^3 + 2) + 5B_1 = x^3 + 10x - 8 \; ,$$
$$H_2 = (x^2 + 2x - 2) + 5A_1 = x^2 + 12x - 7 \; .$$

We use this partial result to find further updates of the initial solution:

3. we solve the Diophantine equation:

$$\frac{F - G_2 \cdot H_2}{5^2} \equiv B_2 \cdot H_2 + A_2 \cdot G_2 \bmod 5$$

   finding the values

$$B_2 = -x - 1; \;\; A_2 = 0; \;\; A_2, B_2 \in \mathbb{Z}_5[x] \; ;$$

4. then we update the solutions

$$G_3 = G_2 + 5^2 B_2 = x^3 + 10x - 8 + 5^2(-x - 1) = x^3 - 15x + 17 \; ,$$
$$H_3 = H_2 \cdot 5^2 \cdot A_2 = x^2 + 12x - 7 + 5^2 \cdot 0 \; .$$

After two steps of iteration we have found G_2 and H_2 which satisfy the following identity:

$$F(x) = (x^2 + 12x - 7) \cdot (x^3 - 15x + 17) \; .$$

Hence G_2 and H_2 represent the exact factorization of $F(x)$.

$$G = (x^3 + 2) \cdot 5^0 + (2x - 2) \cdot 5^1 + (-x - 1) \cdot 5^2 = \sum_{i=0}^{2} B_i 5^i ,$$

$$H = (x^2 + 2x - 2) \cdot 5^0 + (2x - 1) \cdot 5^1 + 0 \cdot 5^2 = \sum_{i=0}^{2} A_i 5^i .$$

Note that at each step we rebuild the p-adic coefficients related to the series expansion of the solution.

We must note that the hypotheses of primality between G_0 and H_0 in Lemma 1 are necessary only to solve the Diophantine equation (1). When F has a specialized form, as in (ii), (iii), and (iv), the equation to be solved is only a modular univariate equation. In these cases, the hypothesis of primality between G_0 and H_0 is no longer needed. In particular, in the numeric case (iv) we need simpler hypotheses, while the algorithm can stay unchanged. Let us show some examples.

Example 3 (The n-th root of a polynomial). Given the following univariate polynomial

$$F(x) = x^4 + 6x^3 - x^2 - 30x + 25$$

we want to compute the square root of $F(x)$ as in (ii). We will solve the equation $\Phi(G) = F - G^2$. If the exact root does not exist we want to compute its approximation. Following the steps previously described we obtain:

1. starting from a suitable initial approximation

$$\Delta_0 = G_0 = x^2 + 3x, \ \ H_0 = 1 ;$$

2. we compute the first-order Taylor series expansion of $\Phi(G)$

$$\Phi(G_0) - 2(G - G_0)\frac{\partial \Phi(G_0)}{\partial G} = -10x^2 - 30x + 25 - 2(G - G_0)(x^2 + 3x) ;$$

3. we solve the equation by letting $(G - G_0) = \Delta_0$

$$-10x^2 - 30x + 25 - 2\Delta_1(x^2 + 3x) \equiv 0 \bmod 5^2$$

   and we find the solution $\Delta_1 = -1$;
4. we update the initial approximation

$$\Delta_0 + \Delta_1 \cdot p = x^2 + 3x + (-1)5 = x^2 + 3x - 5 .$$

In this case we verify that the solution obtained is the exact one:

$$(x^2 + 3x - 5)^2 = F(x) \ .$$

Now we show how the specialization of the Hensel algorithm in the numeric case can provide error-free results.

Example 4. Finding the root of $x^2 - 2$, means to obtain the exact representation of the number $\sqrt{2}$. Let be $F(x) = x^2 - 2$.
Input:
$F(x);\ G = 0;\ H = 0;\ p = 7;\quad G_0 = 0;\ H_0 = 0;\ x_1 = 3;\quad r = 3;$
$(F(3) \equiv 0 \bmod 7)$
$\Delta x_0 = 3;\quad x_1 = 3;$
we want to compute its roots in $\mathbb{Z}_7[x]$. Following the computational steps described in the extended Hensel algorithm we obtain:

1st iteration
$F(x) = F(x_1) + (x - x_1) \cdot F'(x_1),$
$\frac{F(3)}{7} + \frac{x-3}{7} \cdot F'(3) \equiv 0 \bmod 7,$
$1 + \Delta x_1 \cdot 6 \equiv 0 \bmod 7,\quad \Delta x_1 = 1,\quad x_2 = 3 + 1 \cdot 7^1 = 10$

2nd iteration
$\frac{F(10)}{7^2} + \frac{x-10}{7^2} \cdot F'(10) \equiv 0 \bmod 7,$
$2 + \Delta x_2 \cdot 6 \equiv 0 \bmod 7,\quad \Delta x_2 = 2,\quad x_3 = 10 + 2 \cdot 7^2 = 108$

3rd iteration
$\frac{F(108)}{7^3} + \frac{x-108}{7^3} \cdot F'(108) \equiv 0 \bmod 7,$
$6 + \Delta x_2 \cdot 6 \equiv 0 \bmod 7,\quad \Delta x_3 = 6,\quad x_3 = 108 + 6 \cdot 7^3 \in \mathbb{Z}_{7^4}[x].$

At the end of the iterations the algorithm gave

$$x_4 = \sum_{i=0}^{3} \Delta x_i \cdot p^i = 3 + 1 \cdot 7 + 2 \cdot 7^2 + 6 \cdot 7^3$$

and the coefficients of this p-adic expansion $(3\,1\,2\,6)$ represent the mantissa of the Hensel code related to the irrational number $\sqrt{2}$ in the p-adic arithmetic. In fact, in this arithmetic $(3\,1\,2\,6, 0) \times (3\,1\,2\,6, 0) = (2\,0\,0\,0, 0)$, which is the representation of the number 2 in the integer arithmetic.

Example 5. Following a procedure like in the previous example the computation of $F(x) = x^2 - 5$ is briefly shown:
input:
$F(x);\ G = 0;\ H = 0;\ p = 11;\quad G_0 = 0;\ H_0 = 0;\ x_1 = 7;\quad r = 3;$

output:
$x_3 = \Delta x_0 \cdot p^0 + \Delta x_1 \cdot p^1 + \Delta x_2 \cdot p^2 + \Delta x_3 \cdot p^3 = 7 \cdot 11^0 + 6 \cdot 11^1 + 0 \cdot 11^2 + 6 \cdot 11^3$
$= 8059$.

The coefficients of this 11-adic expansion (7 6 0 6) represent the number $\sqrt{5}$ in the p-adic arithmetic. It is simple to verify that, in this arithmetic, $(7\,6\,0\,6, 0) \times (7\,6\,0\,6, 0) = (5\,0\,0\,0, 0)$, which is the representation of the integer number 5.

The importance of dealing with exact results is well known. An example is the following, involving the analytical expression of the n-th Fibonacci number:

$$F_n = \frac{1}{\sqrt{5}} \cdot \left[\left(\frac{1+\sqrt{5}}{2} \right)^n - \left(\frac{1-\sqrt{5}}{2} \right)^n \right] .$$

By using a p-adic representation, we are guaranteed to obtain the exact result, say for F_3, with $p = 11$, $r = 4$, and $\sqrt{5}$ represented by $(7\,6\,0\,6, 0)$.

$$F_3 = \frac{(1\,0\,0\,0, 0)}{(7\,6\,0\,6, 0)} \cdot \left[\left(\frac{(1\,0\,0\,0, 0) + (7\,6\,0\,6, 0)}{(2\,0\,0\,0, 0)} \right)^3 - \left(\frac{(1\,0\,0\,0, 0) - (7\,6\,0\,6, 0)}{(2\,0\,0\,0, 0)} \right)^3 \right] ,$$

$$F_3 = \frac{(1\,0\,0\,0, 0)}{(7\,6\,0\,6, 0)} \cdot \left[\left(\frac{(8\,6\,0\,6, 0)}{(2\,0\,0\,0, 0)} \right)^3 + \left(\frac{(6\,6\,0\,6, 0)}{(2\,0\,0\,0, 0)} \right)^3 \right] ,$$

$$F_3 = \frac{(1\,0\,0\,0, 0)}{(7\,6\,0\,6, 0)} \cdot \left[(4\,3\,0\,3, 0)^3 + (3\,3\,0\,3, 0)^3 \right]$$

where $(4\,3\,0\,3, 0)^3 = (9\,6\,0\,6, 0)$ and $(3\,3\,0\,3, 0)^3 = (5\,6\,0\,6, 0)$

$$F_3 = \frac{(1\,0\,0\,0, 0)}{(7\,6\,0\,6, 0)} \cdot [(9\,6\,0\,6, 0) + (5\,6\,0\,6, 0)] ,$$

$$F_3 = \frac{(3\,2\,1\,1, 0)}{(7\,6\,0\,6, 0)} = (2\,0\,0\,0, 0) .$$

It is worthwhile to stress that the previous computation cannot be performed by the existing systems, without applications of some simplification algorithm. In fact, this is not needed once the proposed p-adic approach is followed.

Example 6. In this example we find, starting from an appropriate initial approximation, a real root of the following polynomial: $F(x) = 6x^2 - 11x + 3$, the zeros of which are $3/2$ and $1/3$. Starting from $x_0 = 2$, $p = 5$, and choosing three iterations for the algorithm, at each iteration we can observe the following output:

$$\Delta x_0 = 2,\ \ \Delta x_1 = 3,\ \ \Delta x_2 = 1,\ \ \Delta x_3 = 3\ ,$$

which are the first four coefficients of the 5-adic expansion of the rational number $1/3$. The other solution, $3/2$, can be found starting from $x_0 = 4$.

5 Conclusion

By means of a uniform representation of mathematical data structures, it is possible to import into a numeric setting the precision guaranteed by the algebraic one. On this basis the integration of numeric and symbolic computation is defined also from a computational point of view.

This integration is founded on the possibility of establishing a precise isomorphism guaranteeing the validity of the approximation method for objects that can be uniformly represented through truncated power series (i.e., numbers and polynomials).

The examples of Sect. 4 show the effectiveness of extending the use of algebraic algorithms to numeric algorithms. Moreover, complex numbers can be represented and treated, as the following example shows.

Example 7. Applying the "integrated" algorithm to the polynomial $x^2 + 1$ with $x_0 = 2$, $p = 5$, and $r = 3$, we obtain the following experimental results:

```
hgen(x^2+1,G,H,x,2,0,0,1,5,3);
```

$$[2, 0, 0]$$

$$[1, 0, 0]$$

$$[2, 0, 0]$$

$$[1, 0, 0]$$

```
summation of the first, 4, coefficients with base, 5, [182, 0, 0]
```

Let us note that $(2\,1\,2\,1, 0) \times (2\,1\,2\,1, 0) = (4\,4\,4\,4, 0)$ which represents the rational number -1.

Finally we discuss some remarks about the choice of initial approximation in the algorithm that we have presented. It must be noted that some problems do still exist. Presently the initial assignments G_0, H_0, or x_0 are chosen by following a heuristic approach. For instance, in each one of the "numeric" examples which have been presented, we have chosen always the smallest base such that an initial approximation does exist.

The analysis of the computational complexity of the algorithm will complete the study of this algorithm, as soon as the authors will finish the very first step of its implementation, i.e., the choice of the suitable base p.

Acknowledgement

This work has been partially supported by CNR under project "Sistemi Informatici e Calcolo Parallelo", grant no. 92.01604.69.

References

Caviness, B. (1986): Computer algebra: past and future. J. Symb. Comput. 2: 217–236.

Char, B. W., Fee, G. J., Geddes, K. O., Gonnet, G. H., Monagan, B. W., Watt, S. M. (1986): A tutorial introduction to Maple. J. Symb. Comput. 2: 171–178.

Colagrossi, A., Limongelli, C., Miola, A. (1997): p-adic arithmetic: a tool for error free computations. In: Miola, A., Temperini, M. (eds.): Advances in the design of symbolic computation systems. Springer, Wien New York, pp. 53–67 (this volume).

Gregory, R., Krishnamurthy, E. (1984): Methods and applications of error-free computation. Springer, New York Berlin Heidelberg.

Hensel, K. (1908): Theorie der algebraischen Zahlen. Teubner, Leipzig.

Lauer, M. (1983): Computing by homomorphic images. In: Buchberger, B., Collins, G. E., Loos, R. (eds.): Computer algebra, symbolic and algebraic computation, 2nd edn. Springer, Wien New York, pp. 139–168.

Limongelli, C., Miola, A. (1990): Abstract specification of numeric and algebraic computation methods. In: Balagurusamy, E., Sushila, B. (eds.): Computer systems and applications, recent trends. Tata McGraw-Hill, New Delhi, pp. 27–35.

Limongelli, C., Temperini, M. (1992): Abstract specification of structures and methods in symbolic mathematical computation. Theor. Comput. Sci. 104: 89–107.

Limongelli, C., Mele, M. B., Regio, M., Temperini, M. (1990): Abstract specification of mathematical structures and methods. In: Miola, A. (ed.): Design and implementation of symbolic computation systems. Springer, Berlin Heidelberg New York Tokyo, pp. 61–70 (Lecture notes in computer science, vol. 429).

Limongelli, C., Miola, A., Temperini, M. (1992): Design and implementation of symbolic computation systems. In: Gaffney, P. W., Houstis, E. N. (eds.): Proceedings IFIP TC2/WG2.5 Working Conference on Programming Environments for High Level Scientific Problem Solving, Karlsruhe, Germany, Sept. 23–27, 1991. North-Holland, Amsterdam, pp. 217–226.

Lipson, J. D. (1976): Newton's method: a great algebraic algorithm. In: Proceedings ACM Symposium on Symbolic and Algebraic Computation, Symsac '76. Association for Computing Machinery, New York, pp. 260–270.

Mascari, G., Miola, A. (1986): On the integration on numeric and algebraic computations. In: Beth, T., Clausen, M. (eds.): Applied algebra, algebraic algorithms and error-correcting codes. Springer, Berlin Heidelberg New York Tokyo, pp. 77–87 (Lecture notes in computer science, vol. 307).

Wirsing, M., Broy, M. (1989): A modular framework for specification and implementation. In: Diaz, J., Orejas, F. (eds.): TAPSOFT '89, vol. 1. Springer, Berlin Heidelberg New York Tokyo, pp. 345–359 (Lecture notes in computer science, vol. 351).

Yun, D. Y. Y. (1974): The Hensel lemma in algebraic manipulation. Ph.D. thesis, Massachusetts Institute of Technology, Cambridge, MA.

p-adic arithmetic: a tool for error-free computations

A. Colagrossi, C. Limongelli, and A. Miola

1 Introduction

In this paper we propose the use of the p-adic arithmetic as a basic computational tool for a symbolic computation system in the framework of the TASSO project. This arithmetic has been chosen for two main reasons.

1. p-adic arithmetic representation provides a unified form to treat numbers and functions by means of truncated power series and it constitutes the mathematical background for the definition of basic abstract data structures for a homogeneous computing environment.
   The problem of the approximation of numbers and functions necessitates an integrated environment allowing a unified representation. We have seen that a unified representation can be obtained when numbers and functions are represented by power series and p-adic analysis offers an appropriate mathematical settlement to handle with power series. Limongelli and Temperini (1997) show how it is possible to treat numbers by truncated power series, as well as the most general p-adic construction methods in an integrated computing environment.
2. p-adic arithmetic is an exact arithmetic and the algebraic bases on which it is founded overcome the problem of the floating point arithmetic, which is essentially due to a lack of algebraic setting. Moreover, the truncated version of this arithmetic that we treat in this paper is suitable to represent numbers in a finite field. This last characteristic belongs to modular arithmetic too (Knuth 1981, Gregory and Krishnamurthy 1984), but the difference is that while modular arithmetic works over the integers, p-adic arithmetic operates on rational numbers. Limongelli (1997) shows the advantages due to the possibility of working directly over the rationals. Moreover, Limongelli and Temperini (1997) and Limongelli (1993a) show that also algebraic numbers are representable in this arithmetic. Limongelli (1997, 1993b) proved the possibility of speeding up the computations via its parallelization from both a theoretical and practical point of view.

Despite its characteristic of modularity and its powerful algebraic properties (completeness of the p-adic metric space; Koblitz 1977), this arithmetic has not received much attentions because of some computational problems, due to the possible lack of the significant digit of the code.

In this paper we show how it is possible to overcome this problem. Actually the only critical point is due to the division operation. We propose a suitable algorithm for the division operation and we show how it is possible to carry out computations with a very low probability of obtaining an actual lack of approximation.

The next section describes the algebraic bases of the proposed arithmetic. Section 3 shows the algebra of the Hensel code set and Sect. 4 is devoted to the treatment of the pseudo-Hensel codes which occur when a significant digit of the code is missed. Section 5 concludes the paper with a description of the possible further steps in order to extend the applicability of this arithmetic.

2 Basic notions

A rational number $\alpha = a/b$, can always be uniquely expressed as

$$\alpha = \frac{c}{d} \cdot p^e$$

where e is an integer, p is a fixed prime number, and c, d, and p are pairwise relatively prime integers. This kind of representation of rational numbers is called the normalized form. Moreover, $\hat{\mathbb{Q}}$ will indicate the set of rational numbers c/d such that $\mathrm{GCD}(d, p) = 1$.

The function

$$\|\cdot\|_p\colon \mathbb{Q} \to \mathbb{R}$$

from the rational numbers $\mathbb{Q}$ to the real numbers $\mathbb{R}$, defined as

$$\|\alpha\|_p = \begin{cases} p^{-e} & \text{if } \alpha \neq 0, \\ 0 & \text{if } \alpha = 0, \end{cases}$$

is a norm on $\mathbb{Q}$ (see Koblitz 1977), called the p-adic norm. On the basis of this p-adic norm, it is possible to define a p-adic metric on $\mathbb{Q}$, such that, given two rational numbers α and β, their distance $d(\alpha, \beta)$ is expressed as:

$$d(\alpha, \beta) = \|\alpha - \beta\|_p \ .$$

Then $(\mathbb{Q}, d)$ is a metric space, $\mathbb{Q}_p$ is the set of equivalence classes of Cauchy sequences in $(\mathbb{Q}, d)$, the system $(\mathbb{Q}, +, \cdot)$ constitutes a field called the field of p-adic numbers, and $(\mathbb{Q}_p, d)$ is a complete metric space.

The main characteristics of the field of p-adic numbers are the following:

- the series

$$\sum_{i=0}^{\infty} p^i$$

converges to $1/(1-p)$ in $(\mathbb{Q}_p, d)$;

– every rational number α can be uniquely expressed in the form:

$$\alpha = \sum_{i=e}^{\infty} a_i p^i; \quad a_i \in \mathbb{Z}_p; \quad e \in \mathbb{Z}; \quad \|\alpha\|_p = p^{-e}; \quad a_e \neq 0 \ , \tag{1}$$

where $\mathbb{Z}$ represents the set of integer numbers.

The p-adic representation of a rational number α is an infinite sequence of digits (the p-adic digits) which are the coefficients of the series given in (1):

$$\alpha = (a_e a_{e-1} \ldots a_{-1} \quad . \quad a_0 a_1 a_2 \ldots) \ .$$

Let us recall that the p-adic expansion of a rational number is periodic. Therefore the p-adic representation can also assume the following form:

$$\alpha = (a_e a_{e-1} \ldots a_{-1} \quad . \quad a_0 \ldots a_{k-m-1}{}' a_{k-m} \ldots a_{k-1} a_k)$$

where the m digits on the right-hand side of the apex constitute the period.

Let us now describe the procedure which computes the p-adic representation of a given rational number α.

p-adic representation of a rational number
Input: p: prime number;
$\alpha = a/b$, represented in its normalized form, $c/d \cdot p^e$;
Output: the coefficients $a_e, a_{e+1}, a_{e+2}, \ldots$ of the p-adic expansion of α;
begin
  $c_1/d_1 := c/d$;
  $i := 0$;
  repeat
    $a_{e+i} := |c_{i+1}/d_{i+1}|_p$;
    $c_{i+2}/d_{i+2} := \frac{1}{p}(c_{i+1}/d_{i+1} - a_{e+i})$;
    $i := i + 1$;
  until the period is detected;
end

Here $|c_i/d_i|_p = |c_i|d_i^{-1}|_p|_p$ is the least non-negative remainder of c_i/d_i mod p.

We note that the hypothesis of primality for p is necessary in order to ensure the existence and the uniqueness of $|d_i^{-1}|_p$. From now on we will consider p a prime number.

Example 1. Let us compute the p-adic expansion of the rational number 3/4, with $p = 5$ (in this case $e = 0$):

$$\alpha = \frac{3}{4} \cdot 5^0, \quad \frac{c_1}{d_1} = \frac{3}{4} \ ;$$

$$a_0 = |c_1/d_1|_p = |3/4|_5 = |3 \cdot |4^{-1}|_5|_5 = |12|_5 = 2 ;$$

$$c_2/d_2 = 1/5(3/4 - 2) = 1/5(-5/4) = -1/4 ;$$

$$a_1 = |c_2/d_2|_p = |-1/4|_5 = ||-1|_5 \cdot |4^{-1}|_5|_5 = 1 ;$$

$$c_3/d_3 = 1/5(-1/4 - 1) = 1/5(-5/4) = -1/4 ;$$

$$a_2 = |c_3/d_3|_p = |-1/4|_5 = 1 .$$

In general, this process will not terminate, but, since we are assuming that α is a rational number, the p-adic expansion will be periodic and we have to continue the detection of the p-adic coefficients until the period is found. In this case the p-adic expansion of the number 3/4 is .211... = .2 '1.

Arithmetic operations on p-adic numbers are carried out, digit by digit, starting from the left-most digit a_e, as in usual base p arithmetic operations.

The division operation on p-adic numbers is performed in a different way w.r.t. usual integer arithmetic.

Starting from the left-most digit of both the dividend and the divisor, we obtain the left-most digit of the quotient, and so on, in a way similar to the other three basic p-adic arithmetic operations.

In order to make automatic the p-adic arithmetic computations, the usual and obvious problem is related to the length of p-adic digit sequence. A natural solution is reached by introducing a finite length p-adic arithmetic on the so-called Hensel codes as we will show below.

Definition 1 (Hensel codes). Given a prime number p, a Hensel code of length r of any rational number $\alpha = (c/d) \cdot p^e$ is a pair

$$(mant_\alpha, exp_\alpha) = (.\ a_0 a_1 \ldots a_{r-1}, e) ,$$

where the left-most r digits and the value e of the related p-adic expansion are called the mantissa and the exponent, respectively. Moreover,

$$\sum_{i=0}^{r-1} a_i \cdot p^i \in \mathbb{Z}_{p^r} .$$

Let $\mathbb{H}_{p,r}$ indicate the set of Hensel codes w.r.t. the prime p and the approximation r and let $H(p, r, \alpha)$ indicate the Hensel code representation of the rational number $\alpha = (a/b) \cdot p^e$ w.r.t. the prime p and the approximation r.

The forward and backward mappings between rational numbers and Hensel codes can then be defined on the basis of the following theorems.

Theorem 1 (Forward mapping). Given a prime p, an integer r and a rational number $\alpha = (c/d) \cdot p^n$, such that $\mathrm{GCD}(c, p) = \mathrm{GCD}(d, p) = 1$, the mantissa $mant_\alpha$ of the code related to the rational number α, is computed by the extended

Euclidean algorithm (EEA) applied to p^r and d as:

$$mant_\alpha \equiv c \cdot y \pmod{p^r}$$

where y is the second output of the EEA.

Proof. See Miola (1984). □

Let us note that the correspondence between the commutative rings $(\hat{\mathbb{Q}}, +, \cdot)$ and $(\mathbb{H}_{p,r}, +, \cdot)$ does not have an inverse, because each Hensel code mantissa $.\, a_0 a_1 \ldots a_{r-1}$ $(= \sum_{i=0}^{r-1} a_i \cdot p^i \in \mathbb{Z}_{p^r})$ in $\mathbb{H}_{p,r}$, is the image of an infinite subset of the rational numbers.

We need to define a suitable subset of $\hat{\mathbb{Q}}$, such that the correspondence between this subset and $\mathbb{H}_{p,r}$ is injective.

Definition 2 (Farey fraction set). The Farey fraction set $\mathbb{F}_{p,r}$ is the subset of $\hat{\mathbb{Q}}$ such that:

$$a/b \in \hat{\mathbb{Q}}\colon \ \mathrm{GCD}(a, b) = 1$$

and

$$0 \le a \le N, \quad 0 < b \le N, \quad N = \left\lfloor \sqrt{\frac{p^r - 1}{2}} \right\rfloor,$$

where $\mathbb{N}$ indicates the set of natural numbers.

$\mathbb{F}_{p,r}$ will also be called the Farey fraction set of order N, as $N = N(p, r)$.

Definition 3. The generalized residue class $\mathbb{Q}_k$ is the subset $\hat{\mathbb{Q}}$ defined as follows:

$$\mathbb{Q}_k = \{a/b \in \hat{\mathbb{Q}} \ \text{ such that } \ |a/b|_{p^r} = k\} \, .$$

From this last definition it follows that

$$\hat{\mathbb{Q}} = \bigcup_{k=0}^{p^r-1} \mathbb{Q}_k \, .$$

Theorem 2. Let N be the largest integer satisfying the inequality

$$2N^2 + 1 \le p^r$$

and let $\mathbb{Q}_k$ contain the order N Farey fraction $x = a/b$. Then x is the only order N Farey fraction in $\mathbb{Q}_k$.

Proof. See Gregory and Krishnamurthy (1984). □

Theorem 3 (Backward mapping). Given a prime p, an integer r, a positive integer $m \leq p^r$ and a rational number $c/d \in \mathbb{F}_{p,r} \subset \hat{\mathbb{Q}}$, let m be the value in $\mathbb{Z}_{p^r}$ of the Hensel code mantissa related to c/d, then the EEA, applied to p^r and m, computes a finite sequence of pairs (x_i, y_i) such that there exists a subscript j for which $x_j/y_j = c/d$.

Proof. See Miola (1984). □

From these considerations we can finally state the following theorem.

Theorem 4. Given a prime p, an approximation r, given an arithmetic operator Φ in $\mathbb{Q}$ and the related arithmetic operator Φ' in $\mathbb{H}_{p,r}$, for any $\alpha_1, \alpha_2 \in \mathbb{Q}$, if

$$\alpha_1 \, \Phi \, \alpha_2 = \alpha_3, \qquad \alpha_3 \in \mathbb{F}_{p,r} \ ,$$

then there exists only one $H(p, r, \alpha_3) \in \mathbb{H}_{p,r}$ such that

$$H(p, r, \alpha_1) \, \Phi' \, H(p, r, \alpha_2) = H(p, r, \alpha_3) \ .$$

On these bases, every computation over $\mathbb{H}_{p,r}$ gives a code which is exactly the image of the rational number given by the corresponding computation over $\hat{\mathbb{Q}}$.

A general schema of computation may consist in mapping on $\mathbb{H}_{p,r}$ the rational numbers given as input to the computation and then of performing the computation over $\mathbb{H}_{p,r}$. However, by Theorem 3, the inverse mapping can be performed only when the expected result belongs to $\mathbb{F}_{p,r}$.

We note that the choice of order of truncation, as well as the choice of the base p, are made in accordance with an a priori estimation of the magnitude of the solution of the problem. In fact, we must identify a suitable set of Farey fractions that contains the rational solution; the choices of p and r are a consequence of this identification.

Such an estimate can be evaluated once a given algorithm is stated for the solution of the problem. This may turn out not to be a simple problem, but such an estimate can be computed on the basis of the number of operations that must be performed in order to reach the rational solution. Let us mention some examples.

- *Arithmetic over the rationals:* Let us consider the computation of a^b, where $a \in \mathbb{Q}$ and $b \in \mathbb{Z}$. The number of bits which are necessary to represent the rational result is $b \cdot \log_2 a$.
- *Algebra of polynomials:* For example, it is easy to compute in advance the maximum coefficient which can be obtained by a polynomial multiplication. In fact, given the polynomials $\sum_{i=0}^{n} a_i \cdot x^i$ and $\sum_{j=0}^{m} b_j \cdot x^j$, if $a = \max\{|a_i|\}_{1 \leq i \leq n}$, $b = \max\{|b_j|\}_{1 \leq j \leq m}$, and $c = \max\{a, b\}$, then the greatest coefficient of the polynomial result is smaller than $\max\{n, m\} \cdot c^2$.

- *Linear algebra:* For example, it is well known that the determinant $D(A)$ of an n-dimensional square matrix A is bounded by $n! \cdot a^n$, where $a = \max\{|a_{i,j}|\}$, $1 \leq i, j \leq n$.

There is also a class of mathematical problems which are particularly well-suited for being solved by p-adic arithmetic: these are problems which are affected either by overflow during the computations or by ill-condition.

3 Operations with Hensel codes

In this section we treat Hensel codes arithmetic and we face the problem of pseudo-Hensel codes manipulation, which essentially consists in a loss of significant digits (i.e., the left-most digits) in a Hensel code. This loss of significant digits does not permit one to execute the division, as stated in literature (Gregory and Krishnamurthy 1984). We will briefly describe the main arithmetic operations, to show that it is possible to overcome this problem, by presenting a new approach both for division computation and for the treatment of the pseudo-Hensel codes. These results, together with the parallelization of p-adic arithmetic, extend its use in a wide class of computing problems.

Let us consider now the arithmetic operations in $\mathbb{H}_{p,r}$.

3.1 Addition

Given two Hensel codes

$$H(p, r, \alpha) = (mant_\alpha, exp_\alpha) \quad \text{and} \quad H(p, r, \beta) = (mant_\beta, exp_\beta) ,$$

first of all we must operate on them, in order to have $exp_\alpha = exp_\beta$, and then perform the addition taking into account that all the operations are carried out from left to right.

Example 2. We want to compute the following addition:

$$\frac{3}{10} + \frac{1}{2}, \text{ in } \mathbb{Z}_{5^4}$$

by choosing $p = 5$ and $r = 4$, we obtain $\mathbb{F}_{5,4} = 17$. The Hensel codes related to $3/10$ and to $1/2$ are respectively:

$$H(4, 5, 3/10) = (.4222, -1) \quad \text{and} \quad H(4, 5, 1/2) = (.3222, 0) .$$

Since the exponents are different, we must normalize the code which has the greater exponent:

$$(.3222, 0) \longrightarrow (.0322, -1) .$$

Now we can carry out the addition between the mantissas:

$$\begin{array}{r} + \ .\,4\,2\,2\,2\ ,\ -1 \\ = \ .\,0\,3\,2\,2\ ,\ -1 \\ \hline .\,4\,0\,0\,0\ ,\ -1 \end{array}$$

The code result $(.4000, -1)$ represents the rational number $4/5$.

Addition behaves as described in Table 1 in which $\mathbb{SH}_{p,r}$ represents the complement of the set $\mathbb{PH}_{p,r}$ with respect to $\mathbb{H}_{p,r}$ (i.e., $\mathbb{H}_{p,r} = \mathbb{PH}_{p,r} \cup \mathbb{SH}_{p,r}$), $\mathbb{PH}_{p,r}$ being the set of the pseudo-Hensel code, given by the following definition:

Definition 4 (Pseudo-Hensel codes). A pseudo-Hensel code ($\mathbb{PH}_{p,r}$) is a code such that $a_0 = \ldots = a_k = 0$, with $0 < k < r - 1$. The order of a pseudo-Hensel code coincides with k.

Table 1. Operational behavior of addition

$+$	$\mathbb{SH}_{p,r}$	$\mathbb{PH}_{p,r}$
$\mathbb{SH}_{p,r}$	$\mathbb{H}_{p,r}$	$\mathbb{SH}_{p,r}$
$\mathbb{PH}_{p,r}$	$\mathbb{SH}_{p,r}$	$\mathbb{PH}_{p,r}$

3.2 Subtraction

In order to perform a subtraction it is sufficient first to compute the complement mod p^r of the minuend and then to carry out the addition. If the minuend is a pseudo-Hensel code, then the subtraction can be carried out in the usual way, without using the complement of the minuend (except in the case when the subtrahend is the Hensel code which represents zero).

Example 3. We want to compute the following subtraction:

$$\frac{3}{4} - \frac{3}{2}, \text{ in } \mathbb{Z}_{5^4} .$$

By choosing $p = 5$ and $r = 4$, we have $\mathbb{F}_{5,4} = 17$. The Hensel codes related to $3/4$ and to $3/2$ are respectively:

$$H(4, 5, 3/4) = (.2\,1\,1\,1\,, 0) \quad \text{and} \quad H(4, 5, 3/2) = (.4\,2\,2\,2\,, 0) .$$

In order to carry out the subtraction, we must get a p-adic unit from the right digit (instead of from the left digit, as usually happens in subtraction between two integer numbers).

$$\begin{array}{rl} - & .\,2\,1\,1\,1\,,\,0 \\ = & .\,4\,2\,2\,2\,,\,0 \\ \hline & .\,3\,3\,3\,3\,,\,0 \end{array}$$

the code result $(.3\,3\,3\,3, 0)$ represents the rational number $-3/4$.

Table 2 shows the operational behavior of subtraction.

Table 2. Operational behavior of subtraction

$-$	$\mathbb{SH}_{p,r}$	$\mathbb{PH}_{p,r}$
$\mathbb{SH}_{p,r}$	$\mathbb{H}_{p,r}$	$\mathbb{SH}_{p,r}$
$\mathbb{PH}_{p,r}$	$\mathbb{SH}_{p,r}$	$\mathbb{PH}_{p,r}$

3.3 Multiplication

In order to perform multiplication, we must operate by multiplying the respective mantissas of the codes, and then we must add their exponents. Also in this case, the code result is truncated to r digits.

Example 4. We want to carry out the following operation:

$$\frac{4}{15} * \frac{5}{2}, \text{ in } \mathbb{Z}_{5^4} .$$

By choosing $p = 5$ and $r = 4$, we have $\mathbb{F}_{5,4} = 17$. The Hensel codes related to $4/15$ and to $5/2$ are respectively:

$$H(4, 5, 4/15) = (.3\,3\,1\,3, -1) \quad \text{and} \quad H(4, 5, 5/2) = (.3\,2\,2\,2, 1) .$$

$$\begin{array}{rl} * & .\,3\,3\,1\,3\,,\,-1 \\ = & .\,3\,2\,2\,2\,,\quad 1 \\ \hline & \;\;4\,0\,0\,0 \\ & \quad\;\,1\,2\,3 \\ & \qquad\;1\,2 \\ & \qquad\;\;\,1 \\ \hline = & .\,4\,1\,3\,1\,,\quad 0 \end{array}$$

The code result $(.4\,1\,3\,1, 0)$ represents the rational number $2/3$.

Table 3 shows the behavior of the multiplication and indicates that the multiplication of the elements of $\mathbb{H}_{p,r}$ is always possible.

Table 3. Operational behavior of multiplication

$*$	$\mathbb{SH}_{p,r}$	$\mathbb{PH}_{p,r}$
$\mathbb{SH}_{p,r}$	$\mathbb{SH}_{p,r}$	$\mathbb{PH}_{p,r}$
$\mathbb{PH}_{p,r}$	$\mathbb{PH}_{p,r}$	$\mathbb{PH}_{p,r}$

3.4 Division

In order to perform division, we must operate by dividing the respective mantissas of the codes, and then we must subtract the respective exponents.

Example 5. We want to carry out the following division:

$$\frac{3}{4}/\frac{6}{5}, \text{ in } \mathbb{Z}_{5^4} .$$

By choosing $p = 5$ and $r = 4$, we have $\mathbb{F}_{5,4} = 17$. The Hensel codes related to 3/4 and to 6/5 are respectively:

$$H(4, 5, 3/4) = (.2\,1\,1\,1\,, 0) \quad \text{and} \quad H(4, 5, 6/5) = (.1\,1\,0\,0\,, -1) .$$

In order to carry out the division, we must compute the inverse mod $p(= 5)$ of the least significative digit of the divisor and then we must multiply it by the left-most digit of the dividend (or of the partial dividend). In this way we will obtain all the digits of the quotient. In this case we must compute $|1^{-1}|_5 = 1$.

$$\begin{array}{cccc|cccc} 2&1&1&1&1&1&0&0\\ \hline 3&2&4&4&2&4&1&3\\ &4&0&0&&&&\\ &1&0&4&&&&\\ &&1&4&&&&\\ &&4&3&&&&\\ &&&3&&&& \end{array}$$

The code result $(.2\,4\,1\,3, 1)$ represents the rational number 5/8.

We must pay particular attention when we operate with elements of $\mathbb{PH}_{p,r}$. In fact, if the first digit of the divisor is zero, we cannot compute the modular inverse as stated in the classical algorithm described in Gregory and Krishnamurthy (1984).

Table 4 indicates when a division results in a pseudo-Hensel code.

Table 4. Operational behavior of division

/	$\mathbb{SH}_{p,r}$	$\mathbb{PH}_{p,r}$
$\mathbb{SH}_{p,r}$	$\mathbb{SH}_{p,r}$	$\mathbb{PH}_{p,r}$
$\mathbb{PH}_{p,r}$	$\mathbb{PH}_{p,r}$	$\mathbb{PH}_{p,r}$

4 Manipulation of pseudo-Hensel codes

Looking at the examples presented above, we note that an addition or subtraction may give a result in which some left-most digits are equal to zero. In this case, we will say that the addition (or subtraction) has generated a pseudo-Hensel code (see Definition 4).

We want to define an algorithm which allows us to overcome the problem of pseudo-Hensel code manipulation and also to decrease the frequency with which they occur during computations, in particular at the end of the computation, when the rational result must be detected. In the following we see when a pseudo-Hensel code can occur.

Example 6. We want to compute the following addition:

$$\frac{13}{15} + \frac{13}{10}, \text{ in } \mathbb{Z}_{5^4} .$$

By choosing: $p = 5$, $r = 4$, we obtain $\mathbb{F}_{5,4} = 17$.

The Hensel codes related to $13/15$ and to $13/10$ are respectively:

$$H(4, 5, 13/15) = (.1\,4\,1\,3\,, -1) \quad \text{and} \quad H(4, 5, 13/10) = (.4\,3\,2\,2\,, -1) .$$

The addition follows:

$$\begin{array}{rl} + & .\,1\;4\;1\;3\;,\;-1 \\ = & .\,4\;3\;2\;2\;,\;-1 \\ \hline & .\,0\;3\;4\;0\;,\;-1 \end{array}$$

In this case one significant digit has been lost. Now, if we apply backward mapping to detect the rational result, we have an error, because the Farey fraction set is decreased: $\mathbb{F}_{5,3} = 7$. In fact, the rational result of this computation is $13/10$, which does not belong to $\mathbb{F}_{5,3}$.

Nevertheless we can carry on with the computation, because the code approximation has not been decreased, but if we want to compute a division in which the dividend belongs to $\mathbb{H}_{p,r}$ and the divisor is a pseudo-Hensel code of order k (with $k < r$), as we have previously observed, we can appropriately manipulate these codes, in order to apply the division algorithm. In the following we will show two algorithms proposed in literature for the manipulation of the pseudo-Hensel codes. The first one was proposed by Dittenberger (1987) and can be summarized by the following steps.

Algorithm Dit87
Input: $q_1 \in \mathbb{PH}_{p,r}$, a pseudo-Hensel code of order k $(1 \leq k \leq r-1)$;
Output: $\bar{q}_1 \in \mathbb{SH}_{p,r-k}$;
1. eliminate the k left-most zeros of the mantissa of q_1;
2. update the exponent of q_1 by adding k to itself;

We note that if we want to carry on with the computation, all the codes involved in the computation must be modified, i.e., their approximation must be decreased by a factor of k. In this way we will eliminate the pseudo-Hensel code, but the approximation of all the codes will be decreased.

The algorithm proposed by Colagrossi and Miola (1987) tries to rebuild the approximation lost by the pseudo-Hensel code occurrence.

Algorithm CM87
Input: $a, b \in \mathbb{SH}_{p,r}$; $q_1 \in \mathbb{PH}_{p,r}$, pseudo-Hensel code of order k $(1 \leq k \leq r-1)$, such that $q_1 = a + b$;
Output: $\bar{q}_1 \in \mathbb{SH}_{p,r}$;
1. apply the inverse mapping to a and to b to obtain the rational numbers related to the codes which have produced the pseudo-Hensel code;
2. apply the direct mapping to these rational numbers, with an approximation equal to $r + k$;
3. carry out the addition: a pseudo-Hensel code of order k is obtained in $\mathbb{PH}_{p,r+k}$;
4. eliminate the first k zeros and update the exponent; now the code will belong to $\mathbb{SH}_{p,r}$;

The latter algorithm presents some inconveniences. While in the former algorithm the approximation of all the codes involved in the computation decreases and could also become zero (in this latter case it is necessary to begin the computation again by doubling at least the code length), in the latter algorithm the computation with the code stops in order to obtain the rational numbers which have generated the pseudo code. But, in this case, we do not know anything about the order of this intermediate result, hence we cannot be certain that backward mapping will give the exact result. Furthermore, this last algorithm is time consuming.

Moreover, let us note that the recovery of the code approximation is necessary only when a division operation with a pseudo code as divisor occurs.

Hence the new proposal consists in proceeding without manipulating the pseudo-Hensel codes, but to manipulate the division algorithm, only when a pseudo-Hensel code occurs as a divisor. In this case, the following algorithm is proposed:

Algorithm Lim92
Input: p, r;
$q_1 \in \mathbb{SH}_{p,r} = (a_0\, a_1\, \dots\, a_{r-1}, 0)$: dividend;
$q_2 \in \mathbb{PH}_{p,r} = (0 \dots 0\, b_k\, \dots\, b_{r-1}, 0)$: divisor
(pseudo-Hensel code of order k);
Output: $q_3 = q_1/q_2 \in \mathbb{PH}_{p,r} = (0 \dots 0\, c_0 \dots c_{r-1-k}, -2k)$;
1. represent the dividend in $\mathbb{H}_{p,r-k}$: $(a_0\, a_1\, \dots\, a_{r-k-1}, 0)$;
2. represent the divisor in $\mathbb{H}_{p,r-k}$: $(b_k\, \dots\, b_{r-1}, k)$;
3. carry out the division (Gregory and Krishnamurthy 1984) in $\mathbb{H}_{p,r-k}$: $(a_0\, a_1\, \dots\, a_{r-k-1}, 0)/(b_k\, \dots\, b_{r-1}, k)$;
4. represent the code result $(c_0 \dots c_{r-1-k}, -2k)$ in $\mathbb{PH}_{p,r}$: $(0 \dots 0\, c_0 \dots c_{r-1-k}, -2k)$. The result will be a pseudo-Hensel code of order k.

Example 7. We want to carry out the following computation:

$$\tfrac{1}{4}/(\tfrac{1}{2} + \tfrac{1}{3}) + \tfrac{1}{25} .$$

We choose, as usual, $p = 5$ and $r = 4$.

$$H(5,4,1/4) = (.4\,3\,3\,3, 0); \quad H(5,4,1/2) = (.3\,2\,2\,2, 0) ;$$
$$H(5,4,1/3) = (.2\,3\,1\,3, 0); \quad H(5,4,1/25) = (.1\,0\,0\,0, -2) ;$$
$$\tfrac{1}{2} + \tfrac{1}{3} = (.3\,2\,2\,2, 0) + (.2\,3\,1\,3, 0) = (.0\,1\,4\,0, 0) ;$$
$$\tfrac{1}{4}/(\tfrac{1}{2} + \tfrac{1}{3}) = (.4\,3\,3\,3, 0)/(.0\,1\,4\,0, 0) .$$

By applying the above described algorithm we obtain:
Input: $q_1 = (.4\,3\,3\,3, 0)$;
$q_2 = (.0\,1\,4\,0, 0)$; (pseudo code of order 1);
Output: $q_3 = (.4\,3\,3\,3, 0)/(.0\,1\,4\,0, 0) = (.0\,4\,2\,2, -2)$;

- we represent the dividend in $H(5,3)$, by truncating the right-most digit of the mantissa:

$$(.4\,3\,3\,3, 0) \longrightarrow (.4\,3\,3, 0) ;$$

- we represent the divisor in $\mathbb{H}_{5,3}$, by truncating the left-most digit of the mantissa. In this case we must update the exponent because the position of the code digits is changed with respect to the power of p:

$$(.0\,1\,4\,0, 0) \longrightarrow (.1\,4\,0, 1) ;$$

– now we carry out the division (Gregory and Krishnamurthy 1984) in $\mathbb{H}_{5,3}$:

$$(.4\,3\,3, 0)/(.1\,4\,0, 1) = (.4\,2\,2, -1) \;;$$

– we represent the code result in $\mathbb{PH}_{5,4}$:

$$(.4\,2\,2, -1) \longrightarrow (.0\,4\,2\,2, -2) \;.$$

Furthermore, $H(5, 4, 1/25) = (.1\,0\,0\,0, -2)$; hence:

$$(.0\,4\,2\,2, -2) + (.1\,0\,0\,0, -2) = (.1\,4\,2\,2, -2) \;.$$

We note that $H(5, 4, 17/50) = (.1\,4\,2\,2, -2)$.

Computing in this way we can avoid the loss of significant digits (on the contrary, in the algorithm Dit87 the approximation order of all the codes involved in the computation decreases) and allows the manipulation of the pseudo-Hensel codes in the same way as the Hensel codes. We do not need to recover the rational numbers related to the Hensel codes which have generated the pseudo code (as does the algorithm CM87). Moreover, the approximation loss can be recovered during the rest of the computation, because, with our algorithm, the code length is not decreased.

5 Conclusions

The occurrences of the pseudo-Hensel codes can be sensibly reduced by taking an appropriate base p and, especially, by parallelizing the arithmetic.

Choosing an appropriate base p, in order to reduce the occurrences of pseudo-Hensel codes, we can note that the probability of finding a leading zero in a code is equal to $1/(p-1)$. The probability of obtaining a leading zero after an addition between two Hensel codes is given by the probability of finding $p-k$ (with $1 \leq k \leq p$) as leading digit of the first code and k as the leading digit of the second code, that is $1/(p-1)^2$. The same occurs for subtraction. From a computational point of view, the best possible choice for p is hence made by taking p to be the greatest prime number less than the maximum integer representable in a memory word.

The parallelization of p-adic arithmetic is not only suitable to reduce the occurrences of the pseudo-Hensel codes, but it speeds up rational number arithmetic especially when big number computations occur. The idea of the parallel p-adic approach (Limongelli 1993b) is to represent each rational number by a Hensel code for several values of p, each representing one homomorphic image. The computations are then performed independently in each image. The unique result has to be constructed out of the results in the images in a recovery step by using the Chinese remainder algorithm. Finally, a backward mapping has to be performed that retrieves the rational number from the Hensel code.

Acknowledgement

This work has been partially supported by CNR under the project "Sistemi Informatici e Calcolo Parallelo", grant no. 92.01604.69.

References

Colagrossi, A., Miola, A. (1987): A normalization algorithm for truncated p-adic arithmetic. In: Irwin, H. J., Stefanelli, R. (eds.): Proceedings of 8th IEEE Symposium on Computation Arithmetic, May 19–21 1987, Villa Olmo, Como, Italy, pp. 212–216.

Dittenberger, K. (1987): An efficient method for exact numerical computation. Diploma thesis, Johannes Kepler University, Linz, Austria.

Gregory, R., Krishnamurthy, E. (1984): Methods and applications of error-free computation. Springer, New York Berlin Heidelberg.

Knuth, D. E. (1981): The art of computer programming, vol. 2, seminumerical algorithms, 2nd edn. Addison-Wesley, Reading, MA.

Koblitz, N. (1977): p-adic numbers, p-adic analysis and Zeta functions. Springer, New York Berlin Heidelberg (Graduate texts in mathematics, vol. 58).

Limongelli, C. (1993a): The integration of symbolic and numeric computation by p-adic construction methods. Ph.D. thesis, University of Rome "La Sapienza", Rome, Italy.

Limongelli, C. (1993b): On an efficient algorithm for big rational number computations by parallel p-adics. J. Symb. Comput. 15: 181–197.

Limongelli, C. (1997): Exact solution of computational problems via parallel p-adic arithmetic. In: Miola, A., Temperini, M. (eds.): Advances in the design of symbolic computation systems. Springer, Wien New York, pp. 68–83 (this volume).

Limongelli, C., Temperini, M. (1997): The uniform representation of mathematical objects by truncated power series. In: Miola, A., Temperini, M. (eds.): Advances in the design of symbolic computation systems. Springer, Wien New York, pp. 32–52 (this volume).

Miola, A. (1984): Algebraic approach to p-adic conversion of rational numbers. Inf. Process. Lett. 18: 167–171.

Exact solution of computational problems via parallel truncated p-adic arithmetic

C. Limongelli

1 Introduction

The aim of the paper is to show the effectiveness of the p-adic arithmetic in scientific computation by selecting and solving problems which manipulates "big" numbers.

By "big" integer number we mean a number which is greater than the maximum integer that can be stored in a word of a given computer. We define a rational number (in reduced form) a/b to be big when a or b (or both) is a big integer.

The reason of our interest in big numbers manipulations is in that they often arise as intermediate or final results in various algebraic algorithms such as Gröbner bases (Buchberger 1965), cylindrical algebraic decomposition (Collins 1975), characteristic sets (Wang 1991), polynomial remainder sequences (Buchberger and Loos 1983), exponentiation and Gaussian elimination (Knuth 1981).

Classical algorithms for rational arithmetic operations (i.e., arbitrary precision rational arithmetic) described by Knuth (1981), are not suitable for big numbers manipulations, due to the cost of GCD algorithm that has to be often applied during computations. Classical modular arithmetic (Knuth 1981) has also one essential lack which depends on the restriction of the set of the numbers on which it works: the integers. When we reformulate a problem stated over the rationals by means of integers, the size of the numbers blows up, producing very slow computations.

We show how it is possible to deal with rational numbers in an efficient way, starting by p-adics and parallelism. We will also see how it is possible to enhance the speed-up of computations for some classes of mathematical problems, by parallel p-adic approach.

The next section presents some definitions necessary to read the paper. The model of p-adic computation will be presented in Sect. 3 in order to introduce the parallel p-adic schema which is described in Sect. 4. Section 5 focuses the attention on the parallelization of the Chinese remainder algorithm (CRA) which is the most expensive step in this model. Two main parallelizations of this algorithm have been proposed and we describe both of them, from a theoretical and a practical point of view. Some experiments are shown in Sect. 6 and Sect. 7 discusses the applicability of this approach and the further steps that have to be done to reach a complete parallelization.

2 Working tools

We skip here basic notions about p-adic arithmetic which can be found in Koblitz (1977), Gregory and Krishnamurthy (1984), and Colagrossi et al. (1997). We just give two main definitions that are necessary for reading the following sections.

Definition 1 (Hensel codes). Given a prime number p, a Hensel code of length r of any rational number $\alpha = (c/d) \cdot p^e$ is a pair

$$(mant_\alpha, exp_\alpha) = (.\, a_0 a_1 \ldots a_{r-1}, e) ,$$

where the left-most r digits and the value e of the related p-adic expansion are called the mantissa and the exponent, respectively. Moreover

$$\sum_{i=0}^{r-1} a_i \cdot p^i \in \mathbb{Z}_{p^r} .$$

Let $\mathbb{H}_{p,r}$ indicate the set of Hensel codes w.r.t. the prime p and the approximation r and let $H(p, r, \alpha)$ indicate the Hensel code representation of the rational number $\alpha = (a/b) \cdot p^e$ w.r.t. the prime p and the approximation r.

Let us consider the problem P involving the computation of a sequence of n arithmetic operations over a given set of m big rational numbers a_i/b_i, $i = 1, \ldots, m$: the p-adic arithmetic can be used to solve this problem once the size of the result is estimated. The result of the computations must belong to the Farey fraction set which is defined as follows.

Definition 2 (Farey fraction set). The Farey fraction set $\mathbb{F}_{p,r}$ is the subset of rational numbers a/b such that:

$$a, b \in \mathbb{N}, \quad 0 \le a \le N, \quad 0 < b \le N, \quad N = \left\lfloor \sqrt{\frac{p^r - 1}{2}} \right\rfloor .$$

Theorem 1 (Computational complexity of forward mapping). The forward mapping algorithm given in Colagrossi et al. (1997: theorem 1), applied to two n-bit numbers, requires $O_B(M(n) \log_2 n)$ time, where $M(n)$ is the time required to multiply two n-bit numbers.

Proof. See Aho et al. (1975). □

Theorem 2 (Computational complexity of backward mapping). The sequential backward mapping algorithm given in Colagrossi et al. (1997: theorem 3), applied to two n-bit numbers, requires $O_B(M(n)\log_2 n)$ time.

Proof. See Aho et al. (1975). □

Let us consider now the algebra on Hensel codes, and the related algorithms, as presented by Gregory and Krishnamurthy (1984); see, however, the corrections by Dittenberger (1987). The following theorem holds.

Theorem 3 (Computational complexity of arithmetic operations). Given two rational numbers α, β and given a prime number p and an approximation r, the computational complexity of the operations $+, -, \cdot, /$ under the bitwise computational model, between $H(p, r, \alpha)$ and $H(p, r, \beta)$ is:
$+\ ,\ -$: $O_B(r\log_2 p)$;
$\cdot\ ,\ /$: $O_B(r^2\log_2 p)$.

Proof. See Limongelli (1987). □

Moreover, in our computations we use the sequential Karatsuba algorithm to multiply integer numbers, whose complexity in terms of time is given as a function of the size z of the input numbers: $O_B(z^{\log_2 3})$.

3 The model of p-adic computation

Before showing the main steps of the algorithm which solves a problem P, let us state the following assumptions:

i. on the basis of the estimated size of the result for P, the values p and r have to be chosen such that this rational result belongs to $\mathbb{F}_{p,r} \subset \mathbb{Q}$;
ii. let a be the size of $\max(a_1, \ldots, a_m, b_1, \ldots, b_m)$, where $q_h = a_h/b_h$, $(h = 1, \ldots, m)$ are the m rational input numbers of the given problem P;
iii. the solution of the problem P requires n arithmetic operations on big rational numbers.

A trivial model for representing the work needed in solving a general problem P, can be sketched in the following few lines. Let us assume that $I_P = \{q_1, \ldots, q_m\}$ is the set of rational numbers given as the input for solving P and let OP_P be the set of the operators on $\mathbb{Q}$ in Assumption (iii): $OP_P = \{op_1, \ldots, op_n\}$. We can define the set C_{P_n} of all the values involved in the computation (including eventually the result of P) as follows:

$$\begin{cases} C_{P_0} = I_P; \\ C_{P_i} = C_{P_{i-1}} \cup q_h\, op_i\, q_{h'}, \qquad q_h, q_{h'} \in C_{P_{i-1}}, \quad h, h' \in \mathbb{N}, \quad i \le n. \end{cases} \tag{1}$$

We can distinguish four main classes of problems, depending on the size of the input data and rational result: the first class is constituted by those problems which have big input numbers and a big output result. Problems in the second class have small input numbers but a big output; problems in the third class start from big input numbers giving small output; problems in the last class have both small input and output data.

We are interested in the first two classes of problems because the other classes can be easily faced by the use of the classic p-adic arithmetic described by Colagrossi et al. (1997).

As far as the first two classes are concerned, we must appropriately choose p, according to the Assumption (i). In this case the term "appropriately" means that we want to avoid overflow during the computations. Hence we fix the base p on the ground of the word-size w of our computer:

$$p \leq 2^{w/2} + 1 \; ; \tag{2}$$

consequently we determine the approximation $\bar{r}$ as follows (see Sect. 2):

$$\bar{r} = \lceil \log_2 p(2N^2 + 1) \rceil \; . \tag{3}$$

Sequential algorithm

Input: p: prime number,
$\bar{r}$: code length, found by relation (3),
$I_P = \{q_h, \; h = 1, \ldots, m\}$: m rational input numbers;
Output: the rational solution $q \in \mathbb{F}_{p,\bar{r}}$;

1.1. Each rational number $q_h \in I_P$; is mapped into the related Hensel code. Let the related set of input data be $I_P^{\mathbb{H}_{p,\bar{r}}} = \{(mant_{q_h}, exp_{q_h}), \; h = 1, \ldots, m\}$, where $(mant_{q_h}, exp_{q_h})$ is the image of the i-th number q_h in the Hensel code set $\mathbb{H}_{p,\bar{r}}$.

1.2. Following the model of computation given in (1), n operations are carried out in $\mathbb{H}_{p,\bar{r}}$, in order to obtain the Hensel code result:

$$(mant_q, exp_q) \; .$$

1.3. The unique rational result is recovered by applying the EEA to the Hensel code result.

In order to analyze the computational complexity of the steps of the algorithm and in order to distinguish between the previous classes of problems, we state further assumptions.

iv.a. $s = \bar{r} \log_2 p$ (the input data are big rational numbers, i.e., P is in the first class);

iv.b. $s < \log_2 p$ (the input data are not big rational numbers, while the intermediate computations involve big numbers and also the output is a big rational number, i.e., P is in the second class);

v. each arithmetic operation in the p-adic domain has maximal computational complexity $O_B(\bar{r}^2 \log_2 p)$, according to Theorem 4.

Theorem 4 (Computational complexity of the sequential algorithm). The total time required to solve P using the sequential algorithm, under Assumptions (i), ..., (v), is:

$$O_B(\max(m(\bar{r} \log_2 p)^{\log_2 3} \log_2(\bar{r} \log_2 p),\ n\,\bar{r}^2 \log_2 p))\ . \qquad (4)$$

Moreover if $n = m^u$, with $u \in \mathbb{N}$, the sequential algorithm takes

$$\begin{cases} O_B(m(\bar{r} \log_2 p)^{\log_2 3} \log(\bar{r} \log_2 p)) & \text{if } u = 1, \\ O_B(m^u(\bar{r}^2 \log_2 p)) & \text{if } u > 1. \end{cases} \qquad (5)$$

Proof. See Limongelli (1993). □

However, it is worth to have a look at theorem 3 enunciated in Colagrossi et al. (1997) which states that the inverse mapping can be performed only when the expected result belongs to $\mathbb{F}_{p,r}$.

Let us note that the truncation order r and the base p are chosen according to a previous estimate of the problem solution. An apt Farey fraction set $\mathbb{F}_{p,r} \subset \mathbb{Q}$, which the rational solution will belong to, must be identified and the appropriate choice for p and r is made accordingly.

Once a given algorithm is stated for the solution of the problem, the related estimate of the solution can be evaluated. This can come out to be a not simple problem, but such an estimate can be computed on the basis of the number of operations that must be performed in order to reach the rational solution. For example, it is well known that the determinant $D(A)$ of a given n-dimensional square matrix A with integer coefficients is bounded by $n! \cdot a^n$, where $a = \max\{|a_{i,j}|\}$, $1 \le i, j \le n$. It is also trivial to estimate the maximum coefficient which raises by polynomial multiplication: given the polynomials $\sum_{i=0}^{n} a_i \cdot x^i$ and $\sum_{j=0}^{m} b_j \cdot x^j$, if $a = \max\{|a_i|\}_{1 \le i \le n}$, $b = \max\{|b_j|\}_{1 \le j \le m}$ and $c = \max\{a, b\}$, then the larger coefficient of the resulting polynomial is bounded by $\max\{n, m\} \cdot c^2$. Also the problem of rational exponentiation is immediately estimated: if we have to compute a/b^c the size of the result will be given by $c \cdot \log_2(\max\{a, b\})$.

4 Parallelization of p-adic computations

In order to develop a parallelization strategy we will refer to a model of shared memory MIMD machine, with a moderate number of processors (about 20 physical processors).

Looking at formulae (4) and (5), we observe that the cost in computing time depends essentially on $\bar{r}^2$, hence a large $\bar{r}$ comes to be time wasteful.

The only way to decrease the approximation $\bar{r}$ should be by increasing the base p. So the design of more efficient algorithms for big rational numbers arithmetic has to deal with balancing the trade-off between a large base p and a reasonable approximation $\bar{r}$.

Here the idea is to reduce the approximation $\bar{r}$ to an approximation $r < \bar{r}$, without increasing the base p (which, in fact, we have limited by the above relation (2)).

In order to exploit the possibilities of parallelization over k processors, we choose several prime numbers p_i, $i = 1, \ldots, k$, such that $g = p_1 \cdot p_2 \cdot \ldots \cdot p_k < p^k$.

On this ground we state the further assumption:

vi. $\text{size}(g) \leq \text{size}(p^k) = k \log_2 p, \quad \text{where } p = \max(p_1, p_2, \ldots, p_k).$

In such a way we carry out the computations in parallel, over the k distinct Hensel code domains $\mathbb{H}_{p_1,r}, \ldots, \mathbb{H}_{p_k,r}$. So the approximation r needed for computing the result of P in $\mathbb{F}_{p,\bar{r}}$ can be obtained by the definition of the Farey fraction set from the relation:

$$(p^k)^r = p^{\bar{r}} ,$$

which implies

$$r = \frac{\bar{r}}{k} . \tag{6}$$

Actually no p_i (base of the computation on the i-th processor) exceeds the limit we imposed above, while the effectively used approximation r is sensibly reduced, making the computations faster.

We assume to have as input a problem P stated over m rational numbers $q_1, \ldots, q_m$. Our goal is to solve it, obtaining a rational result q. We proceed in three main steps (see also Fig. 1):

1. Each rational number q_h $(h = 1 \ldots m)$ is mapped via a prime base p_i to its related Hensel codes $\bar{q}_h^{(i)}$ $(i = 1 \ldots k)$, with a certain truncation order r. All the mappings are done in parallel.

2. The arithmetic expression is evaluated over the $\bar{q}_h^{(i)} \in \mathbb{H}_{p_i,r}$ (sets of Hensel codes). (The image problem is solved in parallel over k finite fields $\mathbb{H}_{p_i,r}$ whose elements are the coefficients of the truncated power series with base p_i and approximation r.)

3. The unique solution $\bar{q} \in \mathbb{H}_{p_1 \cdot \ldots \cdot p_k, r}$ is recovered from the k code results $\bar{q}^{(1)} \in \mathbb{H}_{p_1,r}, \ldots, \bar{q}^{(k)} \in \mathbb{H}_{p_k,r}$, by applying the CRA; then the rational output $\bar{q} \in \mathbb{F}_{p,r}$ is obtained via the backward mapping algorithm. The CRA is parallelized. (Note that the correspondence between $\mathbb{F}_{p,r}$ and $\mathbb{H}_{p,r}$ is injective, hence if the solution does exist in $\mathbb{F}_{p,r}$, then it is unique.)

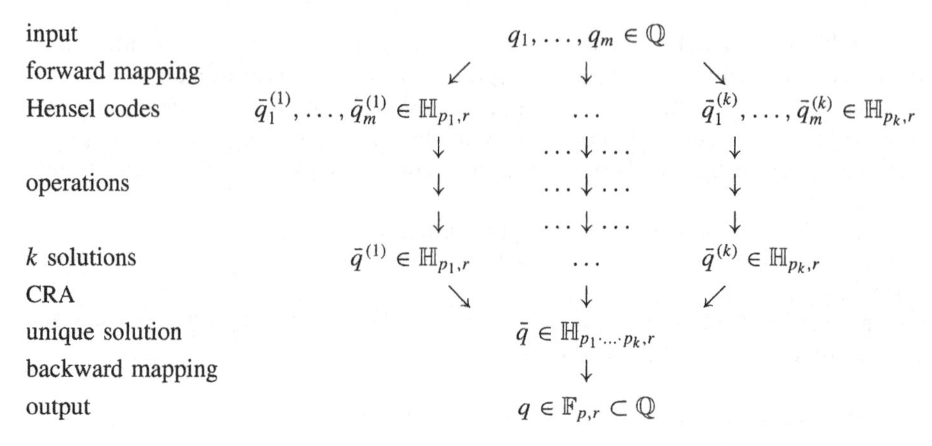

Fig. 1. General schema of parallel p-adic computations

In order to analyze and compare the parallel algorithms, that we are going to define, with the sequential algorithm, we can assume w.l.o.g.:

vii. $r = \log_2 p$. This is a reasonable assumption since if $r > \log_2 p$ then, from Theorem 3 and from Assumption (v), we note that such a larger r would be time wasteful. On the other hand we have to choose a large base p to guarantee the result of the problem $P \in \mathbb{F}_{p,r}$.

Given such a trade-off we can also establish a convenient value for the number k of the processors to be used. Let us note that if $k < r$ then the advantages of parallelization would decrease. Moreover we should actually limit the extent of k to some bound since also a k too great could be unsuitable. Hence, also in order to make less cumbersome the notation in the evaluations of computational complexity, we state the following assumption:

viii. $k = r$.

Moreover, we rewrite the estimate (5) as follows:

$$\begin{cases} O_B(m(r^{3\log_2 3}\log_2 r)) & \text{if } u = 1, \\ O_B(m^u(r^5)) & \text{if } u > 1. \end{cases} \tag{7}$$

On the ground of Assumptions (i)–(viii) the parallel schema of computation is described.

Parallel algorithm
Input: p_i, $i = 1, \dots, k$, number of processors;
r, code length ($r = \bar{r}/k$);
$I_P = \{q_h = a_h/b_h,\ h = 1, \dots, m\}$;
Output: the rational solution $a/b \in \mathbb{F}_{p_1\cdot\ldots\cdot p_k,r}$;

2.1. Each rational number $q_h \in I_P$ is mapped into k Hensel codes, each belonging respectively to $\mathbb{H}_{p_1,r}, \ldots, \mathbb{H}_{p_k,r}$. Let $(mant_{q_{hi}}, exp_{q_{hi}})$ be the image of the h-th number q_h in the i-th Hensel code set.

2.2. The computations are carried out in parallel by the k processors, the i-th processor performs the n operations needed for the solution of P, over $\mathbb{H}_{p_i,r}$, according to the model given in (1), where $i = 1, \ldots, k$ computing the k solutions:

$$(mant_{\gamma_i}, exp_{\gamma_i}) \in C_{P_n} .$$

2.3. a. From the final k results, obtained in the previous step, the unique p-adic result is reconstructed by using the CRA;

b. The EEA is applied to the previous result (like Step 1.3).

Theorem 5 (Computational complexity of the parallel algorithm). The parallel algorithm takes

$$O_B(\max(m\, r^{3\log_2 3} \log_2 r,\; n\, r^3,\; r^{2(1+\log_2 3)})) \tag{8}$$

time, under Assumption (iv.a), and takes

$$O_B(\max(m\, r^{2\log_2 3} \log_2 r,\; n\, r^3,\; r^{2(1+\log_2 3)})) \tag{9}$$

time, under Assumption (iv.b). Moreover, if $n = m^u$, with $u \in \mathbb{N}$, the following relations hold:

$$\begin{cases} O_B(m\, r^{2\log_2 3} \log_2 r) & \text{if } u \le 2, \\ O_B(m^u\, r^3) & \text{if } u > 2; \end{cases} \tag{10}$$

$$\begin{cases} O_B(r^{2(1+\log_2 3)}) & \text{if } u = 1, \\ O_B(m^u\, r^3) & \text{if } u > 1. \end{cases} \tag{11}$$

Proof. See Limongelli (1993). □

The crucial step of this approach is the detection of the unique solution via the CRA.

5 The problem of the recovery step

It is very hard to parallelize the sequential CRA as it is optimized for the sequential case by iterating over the input lists, where in each iteration only a quite cheap computation is necessary. For a more detailed discussion about problems that arise with this parallelized algorithm, see Loidl (1993).

We discuss now three main approaches which have been considered. The first CRA parallelization has been proposed by Colagrossi and Limongelli (1988).

This algorithm starts from the k given Hensel codes

$$(mant_{q_i}, exp_{q_i}) = (q_{0,i} \dots q_{j,i} \dots q_{r-1,i}, exp_{q_i})$$

where $q_{j,i} \in \mathbb{Z}_{p_i}$ represents the j-th digits of the mantissa of the i-th Hensel code. It is based on a "Hensel like" lifting approach and essentially it uses the first $j-1$ ($j = 1, \dots, r$) digits of each Hensel code mantissa $mant_{q_i}$ to obtain the j-th digit δ_j of the g-adic expansion of the final code $\bar{q} \in \mathbb{Z}_g$, where $g = p_1 \cdot p_2 \cdot, \dots, \cdot p_k$ (also called the g-adic result):

$$\bar{q} = \delta_0 + \delta_1 g + \dots + \delta_{r-1} g^{r-1} . \qquad (12)$$

We call this algorithm $\mathcal{S}_{\text{CRA}}$ (stepwise CRA). If we analyze this algorithm described in detail by Colagrossi and Limongelli (1988), we note that the most time consuming operations are the CRA and the computations of the g^j, for $j = 2, \dots, r-1$. We also note that these steps are performed by the use of only one processor.

Therefore, it would be better to base the implementation of the parallel CRA on an idea that contains more inherent parallelism. The following theorem, which is taken from Krishnamurthy (1985), is a good basis for a parallel implementation.

Theorem 6 (Chinese remainder theorem). Let $m_1, \dots, m_k$ be k relatively prime integers > 1. Then for any $s_1, \dots, s_k$ ($s_i < m_i$) there is a unique integer s satisfying

$$s < \prod_{i=1}^{k} m_i =: M$$

and $s_i \equiv s \bmod m_i$; the integer s can be computed using

$$s = \sum_{i=1}^{k} \left(\frac{M}{m_i}\right) s_i T_i \bmod M , \qquad (13)$$

where T_i is the solution of $(\frac{M}{m_i}) T_i \equiv 1 \bmod m_i$.

Limongelli (1993) proposed an efficient algorithm to compute the recovery step which parallelizes the sequential step of the algorithm described above making use of this theorem. In the following we will briefly describe the main steps of this algorithm called $\mathcal{SI}_{\text{CRA}}$ (stepwise improved CRA) and of a new algorithm that is called $\mathcal{P}_{\text{CRA}}$ (pure CRA) that applies the CRA in a more natural way, applying directly Theorem 6. Furthermore, we will compare the complexities of these two algorithms. Firstly let us write input and output specifications which are common to all three algorithms.

CRA specification
Input: $p_1, p_2, \dots, p_k$: k prime modules;

$g = p_1 \cdot p_2 \cdot \ldots \cdot p_k$;
r: code length;
$(mant_{\gamma_i}, exp_{\gamma_i}) = (\gamma_{0,i} \ldots \gamma_{j,i} \ldots \gamma_{r-1,i}, exp_{\gamma_i})$, $i = 1, \ldots, k$:
k Hensel code results;
Output: $\bar{q} = \sum_{j=0}^{r-1} \delta_j \cdot g^j$: the unique code result.

Let us note that we have arbitrarily many virtual processors available that will be automatically mapped on the real processors.

Algorithm $\mathcal{SI}_{\text{CRA}}$ is based on algorithm $\mathcal{S}_{\text{CRA}}$. It essentially applies Theorem 6 at each step of the lifting process.

Algorithm $\mathcal{SI}_{\text{CRA}}$ (Stepwise improved CRA)

1. Apply CRA to the Hensel codes' digit $\gamma_{0,i}$ and moduli p_i, for $i = 1, \ldots, k$, to obtain δ_0;
2. For all j from 1 to $r - 1$ do the following:
   a. Apply the lifting algorithm as described by Colagrossi and Limongelli (1988) to δ_{j-1}, $\gamma_{j,i}$ and moduli p_i to obtain the values $\bar{\gamma}_{j,i}$, for $i = 1, \ldots, k$;
   b. Apply the CRA as it is given in Theorem 6, to the images $\bar{\gamma}_{j,i}$ and to the moduli p_i to obtain δ_j.
   In this case we put $s = \delta_j$, $s_i = \gamma_{j,i}$, $m_i = p_i$, $M = g = p_1 \cdot p_2 \cdot \ldots \cdot p_k$, so we have $(M/m_i) = g_i$ and $T_i = |g_i{}^{-1}|_{p_i}$. Hence the summation (13) becomes:
$$\delta_j = \sum_{i=1}^{k} g_i \cdot \gamma_{j,i} \cdot \bar{g}_i \bmod g \ . \tag{14}$$
3. Transform the digits of the p-adic expansion already obtained, according to the relation (12);

Algorithm $\mathcal{P}_{\text{CRA}}$ applies directly Theorem 6:

Algorithm $\mathcal{P}_{\text{CRA}}$ (Pure CRA)

1. From each Hensel code $(\gamma_i, exp_{\gamma_i})$ compute in $\mathbb{Z}_{p_i^r}$
$$\bar{q}_i = \sum_{j=0}^{r-1} \gamma_{j,i} p_i^j = \gamma_{0,i} + \gamma_{1,i} p_i + \ldots + \gamma_{r-1,i} p_i^{r-1} \ .$$
2. Compute the moduli p_i^r;
3. Apply the CRA as it is given in Theorem 6 to the values $\bar{q}_i = s_i$ and to the moduli $g_i{}^r$ to obtain the unique $\bar{q} \in \mathbb{Z}_{g^r} = s$. Then the summation (13) becomes:
$$\bar{q} = \sum_{i=1}^{k} (g_i)^r \cdot \bar{q}_i \cdot |((g_i)^r)^{-1}|_{g^r} \bmod g^r \tag{15}$$
where $g_i{}^r = (M/m_i)$, $\bar{q}_i = s_i$ and $|((g_i)^r)^{-1}|_{g^r} = T_i$.

Colagrossi and Limongelli (1988) devised an algorithm to perform the recovery step of the unique p-adic solution, with an asymptotic computational complexity of

$$O_B(r^2(k \log_2 p)^{\log_2 3}) , \tag{16}$$

where O_B indicates, as usual, the order of magnitude under the bitwise computational model described by Aho et al. (1975). Limongelli (1993) improved the $\mathcal{S}_{\text{CRA}}$ by parallelizing some steps which occur in the backward mapping of the former algorithm, i.e., CRA and multiplication ($\mathcal{S}_{\text{CRA}}$). The resulting asymptotic computational complexity is

$$O_B(r(k \log_2 p)^{\log_2 3}) . \tag{17}$$

The resulting computational complexity of the algorithm $\mathcal{P}_{\text{CRA}}$ is

$$O_B((r\, k \log_2 p)^{\log_2 3}) . \tag{18}$$

We have to stress that while the $\mathcal{S}_{\text{CRA}}$ step could eventually dominate, in terms of time, the rest of the computation, the $\mathcal{SI}_{\text{CRA}}$ is such that its computational complexity never dominates the asymptotic behavior of the whole algorithm. Then, algorithm $\mathcal{SI}_{\text{CRA}}$ provides a better behavior when it is applied to problems, where the number of operations (assuming that n input rational numbers are involved) needed to reach the solution is at most n^2. Such a class of problems is actually wide, e.g., it contains a large set of operations over matrices and over polynomials.

We have said that the actual problem for parallel computation is the heavy computational complexity of the recovering step, performed by CRA application. Limongelli (1993) showed that the computational complexity of the $\mathcal{SI}_{\text{CRA}}$ does not dominate any more the cost of the parallel computations themselves.

If we want to compare the performance of the parallel algorithm w.r.t. the sequential one, Table 1 shows that, under both Assumptions (iv.a) and (iv.b), when $u > 2$, the sequential algorithm and the parallel algorithms have respectively the complexity

$$O_B(m^u\, r^5) \quad \text{and} \quad O_B(m^u\, r^3) ,$$

where m^u is the number of operations necessary to solve the given expression, and r is the length of the Hensel codes. If we recall Assumptions (vii) and (viii) ($k = r = \log_2 p$) we note that the complexity of the parallel algorithm decreases by a factor $k^2 = r^2$. It is due to the dependency, introduced by (6), of the code length in the sequential algorithm ($\bar{r}$) by a factor k, w.r.t. the code length r adopted by the parallel algorithms.

In Table 1 the related inequalities are summarized (with 1.59 used in place of $\log_2 3$). The result allows to appreciate how better efficiency is reached by the $\mathcal{SI}_{\text{CRA}}$ algorithm w.r.t. the algorithm $\mathcal{S}_{\text{CRA}}$. Moreover, it supports a final evaluation of the general parallel approach w.r.t. the sequential one.

Table 1. Comparison of sequential with parallel algorithms

Seq. Alg.	Par. Alg.	Imp. Par. Alg.	u
Assumption (iv.a)			
$O_B(m(r^{4.77}\log_2 r))$	$= O_B(m(r^{4.77}\log_2 r))$	$= O_B(m(r^{4.77}\log_2 r))$	$u = 1$
$O_B(m^u\, r^5)$	$> O_B(m(r^{4.77}\log_2 r))$	$= O_B(m(r^{4.77}\log_2 r))$	$u = 2$
$O_B(m^u\, r^5)$	$> O_B(m^u\, r^3)$	$= O_B(m^u\, r^3)$	$u > 2$
Assumption (iv.b)			
$O_B(m(r^{4.77}\log_2 r))$	$> O_B(r^{5.18})$	$> O_B(m\, r^{3.18}\log_2 r))$	$u = 1$
$O_B(m^u\, r^5)$	$> O_B((r^{5.18})$	$> O_B(m^u\, r^3)$	$u = 2$
$O_B(m^u\, r^5)$	$> O_B(m^u\, r^3)$	$= O_B(m^u\, r^3)$	$u > 2$

6 Experimental results

In this section we describe the implementation of parallel p-adic arithmetic, to the problem of integer exponentiation. In particular, we focus the attention on the recovery step which is the "weak step" of the resolution algorithm.

The algorithm is written in PACLIB which has been designed at RISC, Linz, especially suitable for the implementation of computer algebra algorithms. This system combines the features of the SACLIB library (Collins et al. 1993, Buchberger et al. 1993) with those of μSystem (Buhr and Stroobosscher 1990). SACLIB is a library of computer algebra algorithms written in C. It is based on a heap management kernel with automatic garbage collections. μSystem is a package supporting light weight processes running under UNIX on shared memory multiprocessors and on workstations.

When using the parallel p-adic approach to parallelize rational number arithmetic, the recovery of a unique result from the results in the homomorphic images is the most time consuming part. Therefore, we compare two algorithms that use two different ways of performing this recovery step. Both variants ($\mathcal{SI}_{\mathrm{CRA}}$ and $\mathcal{P}_{\mathrm{CRA}}$) have been implemented and their dynamic behavior has been evaluated.

In the discussion of the dynamic behavior we concentrate on the case of integer exponentiation and all timings will refer to this case. The reason for choosing this special case is the fact that almost all the implicit parallelism of the p-adic approach lies in the parts that suffice to perform an integer exponentiation (see Fig. 1). Furthermore, it can be easily extended to the general rational case by just adding a final backward mapping step. As we stated in Sect. 3, this problem belongs to the class where the size of input data is definitely smaller than the size of the output data. Hence the recovery step will dominate the computation, but it is useful to stress the recovery step for a detailed comparison of these different versions of the algorithm.

Figure 2a and b shows the dynamic behavior and the utilization of the algorithm based on $\mathcal{SI}_{\mathrm{CRA}}$.

In Fig. 2 the applications of the CRA can be seen as blocks of parallel processes. In the above example, 87654^{500} was computed with $k = 120$. This delivered a Hensel code length of $r = 6$, yielding 5 blocks of parallel processes.

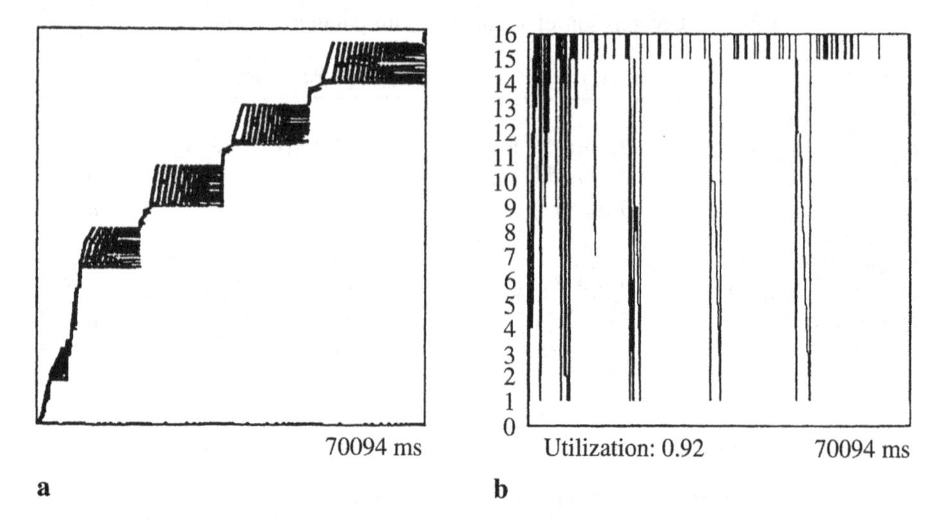

Fig. 2a, b. Dynamic behavior of the algorithm based on $\mathcal{SI}_{\mathrm{CRA}}$. **a** The horizontal axis represents time and the vertical axis represents the parallel tasks. **b** The vertical axis represents the number of active processes.

As for all timings mentioned in this paper, 16 processors have been used. Since k is bigger than the number of processors, many short processes have to be executed. This can be seen from the presence of short lines, instead of long straight lines, inside the blocks. The overhead for starting many parallel processes means an additional loss of efficiency. However, if one uses a smaller k the value for r becomes larger, which means that there are more sequential steps necessary.

Contrary to the previous algorithm, the one based on $\mathcal{P}_{\mathrm{CRA}}$ is founded on the idea of applying the CRA only once on the list of recovered Hensel codes d_i and the moduli m_i. All d_i and m_i can be computed in parallel as their values do not depend on each other. However, the efficiency of this part can be improved by computing blocks of d_i and m_i in each parallel process.

However, the evaluation of d_i and m_i is rather cheap. Timings have shown that these computations require only about 3% of the time needed for the CRA. Therefore, the efficiency of this approach mainly depends on the efficiency of the parallel CRA.

This comparison shows that in practice the new algorithm based on $\mathcal{P}_{\mathrm{CRA}}$, applying the CRA only once on big numbers, is more efficient than the previously known algorithm $\mathcal{SI}_{\mathrm{CRA}}$, which uses a stepwise CRA but operates on smaller numbers. This is due to the fact that the algorithm $\mathcal{P}_{\mathrm{CRA}}$ requires a lower amount of synchronization and therefore is better suited for a parallel execution than algorithm $\mathcal{SI}_{\mathrm{CRA}}$. This fact is not captured in the asymptotic complexity analysis (16) as details like the overhead for the creation of parallel processes are omitted there.

Although only algorithms for rational number exponentiation have been im-

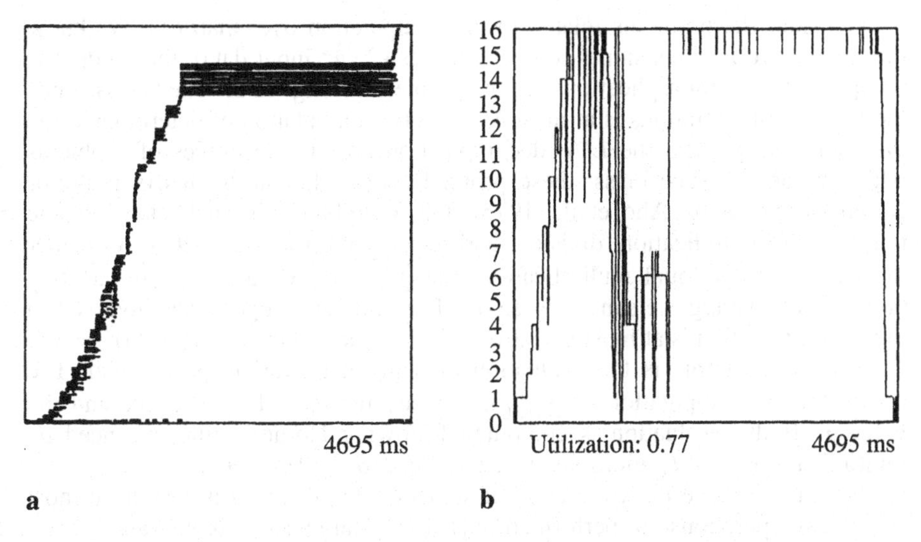

Fig. 3a, b. Dynamic behavior of the algorithm based on $\mathcal{P}_{\mathrm{CRA}}$.

plemented and analyzed, these results hold for general p-adic algorithms as the recovery step is independent of the computation that is performed on the rational numbers.

The big advantage of an algorithm based on this theorem is the fact that it is inherently parallel as all summands in the definition of s (see Eq. (13)) can be evaluated in parallel. Furthermore, the combination of the intermediate results to the final result is rather cheap as only $n - 1$ additions mod M have to be performed.

Figure 3a and b shows the dynamic behavior and the utilization of the algorithm in the evaluation of 87654^{222} with $k = 80$, delivering a Hensel code length of $r = 5$. These input values are different to those for the algorithm based on $\mathcal{SI}_{\mathrm{CRA}}$ in order to see the parallel parts in the execution better.

Contrary to the $\mathcal{SI}_{\mathrm{CRA}}$, there are only twice as much processes as processors necessary in the main step of the algorithm $\mathcal{P}_{\mathrm{CRA}}$. This minimizes the overhead for creating parallel processes. Furthermore, as these processes are completely independent there is no synchronization necessary. This can be seen in Fig. 3 where now the main block consists of straight lines instead of interrupted lines as in Fig. 2.

7 Final remarks

We have shown a parallel model for p-adic computation and we have analyzed the computational complexity of this model, comparing from both a theoretical and practical point of view the algorithms $\mathcal{SI}_{\mathrm{CRA}}$ and $\mathcal{P}_{\mathrm{CRA}}$ which solve the most delicate step of all the computation. As shown in (16)–(18) the $\mathcal{SI}_{\mathrm{CRA}}$ algorithm provides a better behavior when it is exploited for the solutions of problems in which $u \leq 2$. Actually there exists a wide class of such a problems.

Consider the problems related to the computation over matrices. We know that given two n-dimensional square matrices ($2 \cdot n^2$ input data), the product is computed by N^3 multiplications. So the problem belongs to the cited class, since the number of operations is at most $4 \cdot N^4$. The computation of determinant, the matrix inversion, and the *LUP* decomposition are also examples of problems which belong to such class: these problems are reducible to matrix multiplication as shown by Aho et al. (1975). Other problems in such class are the polynomial multiplication, division, and reciprocal (Aho et al. 1975). Consider the case of polynomial multiplication. Let $\sum_{i=0}^{d_a} a_i \cdot x^i$ and $\sum_{i=0}^{d_a} b_i \cdot x^i$ two polynomials having the same degree d_a. The number of operations involved in their multiplication is at most $n = 3(d_a+1)^2 < 4(d_a+1)^2 = (2(d_a+1))^2 = m^2$.

Another example is the evaluation of a given univariate polynomial. The inputs for the computation of $\sum_{i=0}^{d_a} a_i \cdot x^i$ are the $d_a + 1$ coefficients and the value $\bar{x}$. If the evaluation is computed following Horner's rule, we need d_a multiplications and d_a additions, i.e., $n = 2d_a$, $(d_a + 2)^2 = m^2$.

We can also cite the wide class of statistical applications in which, in most cases, few operations are performed over a very large set of input data.

The theoretical analysis of these two algorithms does not agree with the practical one; this is due to the fact that the algorithm $\mathcal{P}_{\mathrm{CRA}}$ requires a smaller amount of synchronization with respect to the algorithm $\mathcal{SI}_{\mathrm{CRA}}$, so its run time behavior is better.

Note that in the implementation of the integer exponentiation a backward mapping of the result is not necessary as it is guaranteed that the resulting number is an integer number. However, other timings have shown that the backward mapping required for the rational case needs about half the time of the overall computation.

Therefore, after having parallelized the main steps of the p-adic approach the sequential backward mapping step becomes the dominant part. As for this backward mapping the application of an EEA is necessary, which is inherently sequential, it is very difficult to parallelize this part. New recent improvements in parallelizing the Euclidean algorithm (Jebelean 1993) can be a starting point for the parallelization of the extended version of this algorithm, namely EEA.

Acknowledgement

This work has been partially supported by CNR under the project "Sistemi Informatici e Calcolo Parallelo", grant no. 92.01604.69.

References

Aho, A. V., Hopcroft, J. E., Ullman, J. D. (1975): The design and analysis of computer algorithms. Addison Wesley, Reading, MA.

Buchberger, B. (1965): Ein Algorithmus zum Auffinden der Basiselemente des Restklassenringes nach einem nulldimensionalen Polynomideal. Ph.D. thesis, University of Innsbruck, Innsbruck, Austria.

Buchberger, B., Loos, R. (1983): Algebraic simplification. In: Buchberger, B., Collins,

G. E., Loos, R. (eds.): Computer algebra, symbolic and algebraic computation, 2nd edn. Springer, Wien New York, pp. 11–43.

Buchberger, B., Collins, G. E., Encarnacion, M., Mandache, A. (1993): A SACLIB 1.1 user's guide. Tech. Rep. 93-19, RISC Linz, Johannes Kepler University, Linz.

Buhr, A., Stroobosscher, R. A. (1990): Providing light-weight concurrency on shared-memory multiprocessors computers running unix. Software Pract. Exper. 20: 929–964.

Colagrossi, A., Limongelli, C. (1988): Big numbers p-adic arithmetic: a parallel approach. In: Mora, T. (ed.): Applied algebra, algebraic algorithms and error-correcting codes. Springer, Berlin Heidelberg New York Tokyo, pp. 169–180 (Lecture notes in computer science, vol. 357).

Colagrossi, A., Limongelli, C., Miola, A. (1997): p-adic arithmetic as a tool to deal with power series. In: Miola, A., Temperini, M. (eds.): Advances in the design of symbolic computation systems. Springer, Wien New York, pp. 53–67 (this volume).

Collins, G. E. (1975): Quantifier elimination for real closed fields by cylindrical algebraic decomposition. In: Brakhage, H. (ed.): Automata theory and formal languages. Springer, Berlin Heidelberg New York, pp. 134–183 (Lecture notes in computer science, vol. 33).

Collins, G. E., Encarnacion, M., Mandache, A., Buchberger, B. (1993): SACLIB user's guide. RISC Linz, Johannes Kepler University, Linz.

Dittenberger, K. (1987): An efficient method for exact numerical computation. Diploma thesis, Johannes Kepler University, Linz, Austria.

Gregory, R., Krishnamurthy, E. (1984): Methods and applications of error-free computation. Springer, New York Berlin Heidelberg.

Jebelean, T. (1993): Improving the multiprecision euclidean algorithm. In: Miola, A. (ed.): Design and implementation of symbolic computation systems. Springer, Berlin Heidelberg New York Tokyo, pp. 45–58 (Lecture notes in computer science, vol. 722).

Knuth, D. E. (1981): The art of computer programming, vol. 2, seminumerical algorithms. 2nd edn. Addison-Wesley, Reading, MA.

Koblitz, N. (1977): p-adic numbers, p-adic analysis and Zeta functions. Springer, New York Berlin Heidelberg.

Krishnamurthy, E. V. (1985): Error-free polynomial matrix computations. Springer, New York Berlin Heidelberg.

Limongelli, C. (1987): Aritmetiche non-standard: implementazione e confronti. Laurea thesis, University of Rome "La Sapienza", Rome, Italy.

Limongelli, C. (1993): On an efficient algorithm for big rational number computations by parallel p-adics. J. Symb. Comput. 15: 181–197.

Loidl, H. W. (1993): A parallel chinese remainder algorithm on a shared memory multiprocessors. Tech. Rep. 93-09, RISC Linz, Johannes Kepler University, Linz.

Wang, D. (1991): On the parallelization of characteristic-set-based algorithms. Tech. Rep. 91-30, RISC Linz, Johannes Kepler University, Linz.

A canonical form guide to symbolic summation

P. Paule and I. Nemes

1 Introduction

Suppose one is interested in a procedure that computes for nonnegative integer input k the values $S(k)$ and $T(k)$ being defined as

$$S(k) := \sum_{j=0}^{k} \frac{2j^2 + 3j - 1}{(j+1)^2(j+2)^2(j+3)} \tag{1}$$

and

$$T(k) := \int_0^k \frac{2x^2 + 3x - 1}{(x+1)^2(x+2)^2(x+3)} \, \mathrm{d}x \ . \tag{2}$$

It is evident that coming up with a procedure for $T(k)$ needs at least some work and nontrivial knowledge, whereas a procedure for computing $S(k)$, a *summation* expression, directly can be read off from the definition. On the other hand, by several reasons it might be desirable to have an equivalent "closed form" expression for $S(k)$, like $-(k+1)/((k+2)^2(k+3))$. Thus one can derive procedures of different kinds for representing the summation sequence $(S(k))_{k \geq 0}$.

For different purposes different representations of sums might be desired, e.g., for analysis of algorithms, for combinatorial enumeration, for number theoretic investigations, etc. These representations usually can be obtained by various techniques, classical as well as computer-assisted ones, ranging from ad hoc argumentation to systematic use of theory, and from table look-up to fully computerized derivation. For an excellent source providing lots of instructive examples see Graham et al. (1994). One could ask in which mathematical or algorithmic aspects these techniques differ, or for a precise description of what is understood to be in "closed form". It is the objective of this article to provide a framework, on a more or less informal level, together with illustrating examples, how these questions could be approached by using the notions of "simplification" and "canonical simplification" as described in detail, e.g., in the classical survey paper by Buchberger and Loos (1983). There one also finds an extensive bibliography of relevant articles. For surveys concerning the algorithmic and technical aspects of symbolic summation, see, e.g., Lafon (1983) or Graham et al. (1994). For a discussion of recent developments, see Paule and Strehl (1995).

In the remainder of this section we give a brief introduction to the indefinite summation problem, recall what is understood by (canonical) simplification, and introduce sequence-domains as fundamental sets of expressions we deal with.

In Sect. 2 we consider standard operations that turn sequence-domains into (admissible) a-algebras. We focus on those types of representation that are used most frequently: "explicit", generating function, and recurrence representation. Because of their prominent role in applications, special emphasis is put on domains representing hypergeometric sequences. Nevertheless, the framework is kept sufficiently general in order to fit to non-hypergeometric instances and to other types of representation as well.

In Sect. 3 the indefinite summation problem is described as finding a sum-quantifier free representation. This is done via embedding an admissible sequence-domain in two free term algebras. One of those, where the solution is searched for, has a sum-quantifier free signature. With respect to a given canonical simplifier *CS* the problem is reformulated as the problem of extending *CS* to sum expressions. In this framework, which can be carried over to definite summation as well, the notion of "closed form" takes on a certain precise meaning.

In Sect. 4 certain aspects of generating function representation are treated, focusing on hypergeometric and, more generally, on P-finite sequences.

In Sect. 5 several illustrating examples are given. They range from an American Mathematical Monthly problem to a proof of a recent monotonicity conjecture of Knuth. Special emphasis is put on showing the importance of being sufficiently flexible in choosing appropriate representation domains under consideration of the conversion problem.

1.1 Introductory examples

Let **N** be the set of nonnegative integers. By **F** we denote a finite field extension of the rational number field. In order to keep technicalities down to a minimum, it will be convenient to take a field, despite the fact that a large part of the following can be formulated also for other algebraic structures as rings, etc. By $\mathbf{F}^{\mathbf{N}}$ we understand the class of all **F**-sequences.

Informally the problem of indefinite summation can be described as follows. Given an **F**-sequence $(t_j)_{j\geq 0}$, find a "closed form" representation of the **F**-sequence $(s_k)_{k\geq 0}$ defined by

$$s_k = \sum_{j=0}^{k} t_j \ .$$

For better illustration we present a simple binomial summation. The binomial coefficients are defined as $\binom{a}{j} := \frac{a^{\underline{j}}}{j!}$, where $a^{\underline{0}} = 1$, and $a^{\underline{j}} = a(a-1)\dots(a-j+1)$ for $j \geq 1$. Let $t_j = (-1)^j\binom{a}{j}$ and consider the sum sequence of $(t_j)_{j\geq 0}$. If s_j denotes $\sum_{i=0}^{j} t_i$ we have

$$s_j = (-1)^j\binom{a-1}{j} , \tag{3}$$

which certainly can be considered as a "closed form" representation. The proof of (3) is immediate from $s_j - s_{j-1} = t_j$, for $j \geq 1$, which is equivalent to the standard recurrence of the binomial coefficients.

Often this situation is rephrased in terms of "identities". For instance, the example above can be found in Gould's (1972) table as entry (1.5) in the form

$$\sum_{j=0}^{k}(-1)^j\binom{a}{j} = (-1)^k\binom{a-1}{k}. \tag{4}$$

An alternative way to state identity (4) is in terms of generating functions:

$$\frac{1}{1-x}\cdot(1-x)^a = (1-x)^{a-1}.$$

By taking the coefficient of x^k on both sides the equivalence becomes evident. An excellent reference for generating functions is Wilf (1990).

A further elementary example concerns summation of polynomial sequences like

$$\sum_{j=0}^{k} j = \frac{k(k+1)}{2}. \tag{5}$$

Besides generating functions, another alternative to state summation identities is in terms of recurrences. For instance, identity (5) is equivalent to saying that both sides are solutions of the recurrence

$$f_{k+1} - f_k = k+1$$

and coincide at the initial value $k = 0$.

It is well known that the sum sequence of a polynomial sequence again is polynomial. For summing rational sequences this closure property does not hold anymore. For instance, consider

$$\sum_{j=0}^{k}\frac{1}{j+1} =: h_k,$$

the so-called *harmonic number* sequence. One can prove, e.g., by applying Gosper's algorithm, that there exists no rational function $r(x) \in \mathbf{F}(x)$ such that $h_k = r(k)$ for all $k \geq 0$. This observation might suggest to view the sum-representation itself as a "closed form".

In instances like the ones above and in most applications the summation problem and the notion of "closed form" intuitively are clear from the context or determined by practical demands (complexity considerations etc.). But especially for algorithmic treatment a more refined analysis of the problem is desirable. For instance, as we shall see below, in certain contexts the harmonic number sequence indeed finds a closed form representation. In general, the various symbolic summation methods differ in their representation domains and/or by the way they change from one type of representation to another. It is the intention of this paper to provide a problem specification that brings all these aspects under one umbrella. The guiding principle behind the setting is the fundamental notion of (canonical) simplification.

1.2 Simplification

The following general description of the problem of (canonical) simplification is taken from Buchberger and Loos (1983). Let U be a class of (linguistic) objects called "expressions" (e.g., terms of certain types, logical formulae, or programs) and let $\sim$ be an equivalence relation on U (e.g., functional equivalence, equality derivable from axioms, etc.). Then the problem of canonical simplification consists in finding an effective procedure, a "canonical simplifier", CS: $U \to U$ such that for all $s, t \in U$:

$$CS(t) \sim t \quad \text{and} \quad s \sim t \Rightarrow CS(s) \equiv CS(t) .$$

More generally, the problem of obtaining equivalent but simpler objects consists in finding an effective procedure, a "simplifier", SI: $U \to U$ such that for all $t \in U$:

$$SI(t) \sim t \quad \text{and} \quad SI(t) \leq t .$$

Here, $\leq$ is the concept of simplicity one is interested in.

The linguistic objects represent mathematical objects. For instance, the elementary arithmetical expressions $j(j+1)/2$ and $j^2/2 + j/2$ are not identical as linguistic objects, i.e., $j(j+1)/2 \not\equiv j^2/2 + j/2$. They are equivalent, i.e., $j(j+1)/2 \sim j^2/2+j/2$, in the sense that they represent the same polynomial in $\mathbf{F}[j]$. Another possibility is to interpret them as equivalent representations of the triangular-number sequence $(0, 1, 3, 6, \ldots)$. It is the latter, i.e., the $\mathbf{F}$-sequence interpretation that will be discussed in this paper. For the sake of simplicity we shall consider only settings in which the following holds. Whenever two expressions will be functionally equivalent they also will satisfy equality derivable from axioms, and vice versa. Therefore equivalence "$\sim$" will be written with the usual symbol "$=$", otherwise it will be specified explicitly.

The way how sequences are represented plays a fundamental role in the context of symbolic summation. Thus, first of all we specify which kind of expressions will be taken for that. A *sequence-domain* is a pair $(X, eval)$, where X is a set of expressions on which an evaluation function

$$eval\colon \mathbf{N} \times X \to \mathbf{F}, \qquad (n, s) \mapsto s[n]$$

is defined. We think of *eval* as an effective procedure that computes $s[n]$ for a given expression s in a finite number of steps.

Accordingly, $s, t \in X$ are equivalent, this means $s = t$, if $s[n] = t[n]$ for all $n \in \mathbf{N}$.

In the literature three different sequence-domains are used most frequently, namely:

- elementary (or "explicit"),
- generating function,
- recurrence (or "difference equation")

representation. The domains will be abbreviated correspondingly by *EleEx*, *GenFu*, and *RecEq*. Instead of providing a concise definition, we present an illustrating example.

Example. Consider the constant sequence $a := (a_j)_{j\geq 0}$ with $a_j = 1$ for all $j \geq 0$, and the sequence of positive integers $b := (j+1)_{j\geq 0}$.

a. Then a can be represented alternatively by its constant value $s_1 := 1 \in$ *EleEx*, or, by its ordinary generating function $s_2 := 1/(1-x) \in$ *GenFu*, or, by the recurrence $s_3 := (E - I, 0; 1) \in$ *RecEq*. The recurrence expression is equivalent to

$$a_{j+1} - a_j = 0 \quad \text{for } j \geq 0,\ a_0 = 1 ,$$

where $E - I$ stands for the shift minus identity operator (applied to a_j), the second entry in $(E - I, 0; 1)$ stands for the homogeneous part 0, and the third entity for the initial value 1. The corresponding evaluation functions are obvious for any $j \in \mathbf{N}$, i.e., $s_1[j]$ is obtained via (trivial) substitution, $s_2[j]$ is obtained by a procedure computing the j-th coefficient in the power-series expansion of $1/(1-x)$, and $s_3[j]$ is obtained from a recursive procedure following the steps $s_3[j+1] = s_3[j]$ with initial value $s_3[0] = 1$.

b. Similarly, b can be represented by a corresponding elementary term $t_1 := j + 1 \in$ *EleEx* (j being a free variable), or, by its generating function $t_2 := 1/(1-x)^2 \in$ *GenFu*, or, by the recurrence $t_3 := (E - I, 1; 1) \in$ *RecEq*. The corresponding evaluation functions are the same as above, e.g., $t_1[j]$ is obtained via (polynomial) substitution, a.s.o.

c. We want to note that in general the representations are not unique. For instance, if we admit binomials as elementary expressions, one has

$$s_1 = \binom{j}{j} \quad \text{or} \quad t_1 = \binom{j+1}{1} .$$

If one takes formal power series or elements from a differential field (with D as the derivation operator) as generating function expressions, one has

$$s_2 = \sum_{n=0}^{\infty} x^n = \exp(-\log(1-x)) \quad \text{or} \quad t_2 = D\frac{1}{1-x} .$$

Also recurrence expressions need not be unique, e.g.,

$$t_3 = ((j+1)E - (j+2)I, 0; 1) ,$$

a.s.o.

d. It should be also noted that the various representations not only differ with respect to their specific form, but also in the corresponding evaluation functions.

For instance, for $n \in \mathbf{N}$ the evaluation $t_1[n]$ is a trivial one step substitution, whereas $t_3[n]$ needs $n-1$ (simple) addition operations.

Remark. Besides these types of representation, integral representation plays an important role especially in connection with special function identities. For a comprehensive account of treating combinatorial sums with this technique we refer the interested reader to Egorychev (1984).

Roughly speaking, the problem of symbolic summation can be viewed as the problem of finding a "simpler" expression for the sum in question. One alternative to define "simplicity" of an expression from a sequence-domain would be with respect to the evaluation function *eval*. In our approach, i.e., searching for a "sum-quantifier free representation" (see Sect. 3), we follow another alternative that is very close to definitions one finds in literature. However, being guided by the unifying concepts of simplification and canonical simplification, the emphasis is put on keeping the defining notions as flexible as possible. This way it is possible to bring classic and recent summation methods under one umbrella (see Sect. 5). Also, it should not be too difficult to modify the given approach, in order to match needs that are not covered explicitly. For instance, one could consider modifications with respect to simplicity of evaluation functions, as indicated above.

2 Sequence-domain algebras

In this section we set up the stage for defining the problem of symbolic summation in Sect. 3. After considering operations on sequences or on expressions from sequence domains, respectively, embedding into two term-algebras, one being sum-quantifier free, is introduced. Certainly, various alternative formulations could be chosen, nevertheless the presented one seems to be close to practical demands.

2.1 Operations on sequences

On $\mathbf{F}$-sequences several standard operations can be considered such as: (componentwise) addition $\oplus$, (Cauchy) product $*$, (componentwise) product $\odot$ (also called Hadamard-product), and interlacing *Interlace* (see Sect. 5.2). There are also unary operations like the L-transform, (forward) shift E, (forward) differencing Δ, sum Σ, and multisection *Multisect*. For $f = (f_j)_{j\geq 0}$ the shift means $E(f) := (g_j)_{j\geq 0}$ with $g_j = f_{j+1}$, and differencing $\Delta(f) := (g_j)_{j\geq 0}$ with $g_j = f_{j+1} - f_j$, whereas the sum-transform and the $*$ operation involve a sum-quantifier being defined as

$$\Sigma(f) := (g_k)_{k\geq 0} \quad \text{where } g_k = \sum_{j=0}^{k} f_j \ ,$$

and, if $\bar{f} = (\bar{f}_j)_{j\geq 0}$,

$$f * \bar{f} := (h_k)_{k\geq 0} \quad \text{where } h_k = \sum_{j=0}^{k} f_j \bar{f}_{k-j} \ .$$

By $L(f) := (g_j)_{j\geq 0}$ we understand the transformation to the constant initial value sequence with $g_j = f_0$ for all $j \geq 0$. These operations satisfy several equations. For instance, it is straightforward that for all $f \in \mathbf{F}^{\mathbf{N}}$ we have

$$\Delta(\Sigma(f)) = E(f) \quad \text{and} \quad \Sigma(\Delta(f)) = E(f) \oplus (-L(f)) \ .$$

The latter equality corresponds to the *telescoping sum property*, i.e.,

$$\sum_{j=0}^{k} (f_{j+1} - f_j) = f_{k+1} - f_0 \tag{6}$$

for all $k \geq 0$.

In the following a will denote an arity-function (signature) with respect to some or all of these operations. For instance, $a = ((\oplus, 2), (E, 1))$ maps $\oplus$ and E to the arities 2 and 1, respectively. We will shortly write $a = \{\oplus, E\}$.

An arity-function a will be called *sum-quantifier free*, if the operations $\sum$ and $*$ are not included in a. We fix σ to be the ("standard") signature

$$\sigma := \{\oplus, \odot, *, L, E, \Delta, \Sigma\} \ ,$$

and τ as the ("standard") sum-quantifier free subset

$$\tau := \{\oplus, \odot, L, E, \Delta\} \ .$$

We want to note that for various needs the "standard" signatures might be extended by binary operations like *Interlace*, *Multisect*, or by unary operations like scalar-multiplication, or differentiation (as for generating function representation).

Evidently $\mathbf{F}^{\mathbf{N}}$ is an a-algebra with $a = \sigma$. (For precise definitions of notions like a-algebra, term algebra, generating set, etc., see, e.g., Buchberger and Loos 1983.)

Let X be a sequence-domain, i.e., it is given an evaluation function such that for each expression $s \in X$ we have $s[n] \in \mathbf{F}$ for all $n \in \mathbf{N}$.

Let the term algebra $T = Term(X, a)$ be the free a-algebra with generating set X. If U is an a-algebra with $a = \{\oplus, \odot, \ldots\}$, we denote by $\oplus_U, \odot_U$, etc., the corresponding operations in U. For operations in the a-algebra $\mathbf{F}^{\mathbf{N}}$ we use the same symbols $\oplus, \odot$, etc., without indices.

Let the map v: $T \to \mathbf{F}^{\mathbf{N}}$, $s \mapsto v(s)$ be defined as follows: for $s \in X$ define $v(s) := (s[j])_{j\geq 0}$, and $v(t)$ for all $t \in T$ by the corresponding homomorphic extension. (Such a homomorphism v is called an assignment in $\mathbf{F}^{\mathbf{N}}$, cf. Buchberger and Loos 1983.) For instance, for $s, t \in X$ with respect to the (unary) operation $\sum$ and the (binary) Hadamard-product we have

$$v(\sum(s)) = \Big(\sum_{j=0}^{k} s[j]\Big)_{k\geq 0}$$

and

$$v(s \odot t) = (s[j]t[j])_{j\geq 0} ,$$

respectively.

2.2 Admissible sequence-domains

For the treatment of sum-expressions $\Sigma(f)$, where $f \in \mathbf{F}^{\mathbf{N}}$, one first has to fix an appropriate sequence-domain X. Let be $t \in X$ such that $v(t) = f$. In view of the telescoping relation (6), we impose that X is an *admissible* domain, this means an a-algebra with $\{L, E, \Delta\} \subseteq a$. More precisely, the subset relation means that X at least is equipped with unary operations $L_X, E_X, \Delta_X \colon X \to X$ such that $v(L_X(s)) = L(v(s))$, $v(E_X(s)) = E(v(s))$, and $v(\Delta_X(s)) = \Delta(v(s))$ for all $s \in X$. Let A be a set of rules, e.g., monoid or commutative ring axioms, that induce (via "equational calculus", see Buchberger and Loos 1983) an equivalence relation $=_A$ on X such that $s_1 =_A s_2$ iff $s_1 = s_2$. (The latter equality means functional equivalence, i.e., $v(s_1) = v(s_2)$.)

Most of the notions introduced will be illustrated according sequences of hypergeometric type which play a prominent role in various applications. A sequence is called *hypergeometric* if there exists a rational function $r \in \mathbf{F}(x)$ such that

$$\frac{f_{j+1}}{f_j} = r(j) \quad \text{for all} \quad j \in \mathbf{N} . \tag{7}$$

Remark. Note that this definition excludes sequences containing zeros. A careful treatment of more general sequences like $\binom{n}{k}(k-2)$, has been done by Schorn (1995).

Most of the Taylor-series expansions arising in applications are (infinite) sums over hypergeometric sequences. With respect to symbolic summation it is important to note that this class provides a canonical form representation based on the order 1 recursion (7), which can be described uniquely by a canonical form of r together with the initial value f_0. Most of the binomial summations are (finite) sums over hypergeometric sequences. The drawback with binomial coefficient expressions comes from the observation that they arise in various "disguised" forms, e.g.,

$$\binom{n}{j} = \frac{n}{j}\binom{n-1}{j-1} = \frac{j+1}{n-j}\binom{n}{j+1} = (-1)^j\binom{j-n-1}{j} = \dots$$

Remark. With respect to table look-up methodology the treatment of binomial sums within the theory of hypergeometric functions (see, e.g., Graham et al. 1994) in most instances turns out to be far superior to applying techniques for direct manipulation of binomial expressions (as described, e.g., in Knuth 1973). This is mainly due to the fact mentioned above, i.e., hypergeometric notation provides a *canonical form* for binomial summand sequences. From practical point of view, compared to Gould's (1972) list of 500 binomial identities the most important hypergeometric identities only fill a few pages, e.g., the appendix of Slater (1966).

For better illustration we give several examples of representations of hypergeometric sequences, for instance, as elementary expressions $HypEx \subseteq EleEx$. For representation as generating functions $HypFu \subseteq GenFu$, as recurrences $HypEq \subseteq RecEq$, and as elements of certain difference field extensions, see Sect. 4. All these sequence-domains X are a-algebras of type $(X, \{\odot, L, E, \Delta\})$, this means they are closed (at least) under Hadamard-product etc. Closure with respect to $\oplus$ only can be achieved in suitable extensions, because the addition of two hypergeometric sequences does not need to be hypergeometric again. A simple example is provided by $(a_j + b_j)_{j\geq 0}$, where $a_j = 2^j$ and $b_j = 1$.

Example. Let $a^{\bar{k}} := a(a+1)\dots(a+k-1)$ be the rising factorial for $k \geq 1$ and $a^{\bar{0}} := 1$. Then from the recursion (7) it is immediate that hypergeometric sequences can be brought into the form

$$\frac{a_1^{\bar{j}} \dots a_m^{\bar{j}}}{b_1^{\bar{j}} \dots b_n^{\bar{j}}} \, rat(j)\, c^j \ , \tag{8}$$

where $m, n \in \mathbf{N}$, $a_k, b_l, c \in \mathbf{F}$ and $rat \in \mathbf{F}(x)$. Let $HypEx$ be the set of expressions of this type where j is a free variable, and *eval* the obvious evaluation which is nothing but nonnegative integer substitution for j. Then the sequence-domain $X := (HypEx, eval)$ is an a-algebra of type $(X, \{\odot, L, E, \Delta\})$. Hadamard-product is defined as usual, and if $t \in HypEx$ is an expression of the form (8) then $L_X(t) := rat(0)$ (i.e., $m = n = 0$ and $c = 1$),

$$E_X(t) := \frac{a_1^{\bar{j}} \dots a_m^{\bar{j}}}{b_1^{\bar{j}} \dots b_n^{\bar{j}}} \, Rat(j)\, c^j \ ,$$

where $Rat(j) = c \cdot rat(j+1)(a_1 + j)\dots(a_m + j)/((b_1 + j)\dots(b_n + j))$, and

$$\Delta_X(t) := \frac{a_1^{\bar{j}} \dots a_m^{\bar{j}}}{b_1^{\bar{j}} \dots b_n^{\bar{j}}} \, \overline{Rat}(j)\, c^j \ ,$$

where $\overline{Rat}(j) = Rat(j) - rat(j)$.

2.3 Term algebra embeddings

Given an admissible sequence-domain X (i.e., X is an a-algebra with $\{L, E, \Delta\} \subseteq a$) together with an axiom set A inducing equivalence as above. Then the a-algebra X finds a canonical homomorphic embedding into the two free term algebras $T := Term(X, \tau)$ and $S := Term(X, \sigma)$ generated by X. Note that the signature a of X might contain also $*$ or Σ, hence the embedding into T is understood by restriction to the sum-quantifier free operations in a.

As A for X, let A_τ, resp. A_σ, be the corresponding axiom sets inducing equivalence, e.g., commutative ring axioms like

$$(r \odot s) \odot t = r \odot (s \odot t), \quad (r \oplus s) \odot t = (r \odot t) \oplus (s \odot t) ,$$

or difference ring axioms like

$$E(s \odot t) = E(s) \odot E(t), \quad \Delta(s \oplus t) = \Delta(s) \oplus \Delta(t) ,$$

or convolution axioms like

$$\Sigma(s) = s * \mathbf{1} = \mathbf{1} * s$$

(here **1** denotes the corresponding expression for the constant sequence consisting of 1's only), or the "telescoping" axioms

$$\Delta(\Sigma(s)) = E(s), \quad \Sigma(\Delta(s)) = E(s) \oplus (-L(s)) .$$

The last equation is one of the reasons why, with respect to summation, it is of advantage to work in the extension T of X. For example, for a given $t \in X$ one might be able to find an $s \in X$ such that $t = \Delta(s)$, which in principle would solve the summation problem as $\Sigma(t) = E(s) \oplus (-L(s))$. But, despite the fact that $E(s), (-L(s)) \in X$, it is not necessarily true that their addition is in X, which happens, e.g., if $X = HypEx$.

Induced by the homomorphic embeddings we get additional equivalence axioms. More precisely, for unary $\omega \in a$ and binary $\circ \in a$, we define for all $s, t \in X$:

$$\omega_S(s) = \omega_X(s) \quad \text{and} \quad s \circ_S t = s \circ_X t ,$$

and analogously for the sum-quantifier free case

$$\omega_T(s) = \omega_X(s) \quad \text{and} \quad s \circ_T t = s \circ_X t .$$

If $A_{X \to T}$, resp. $A_{X \to S}$, denotes the set of these axioms, we consider equivalence on T, resp. S, as being induced by the set $A_T := A_\tau \cup A_{X \to T}$, resp. $A_S := A_\sigma \cup A_{X \to S}$.

Evidently, T is a subalgebra of S, and $A_T \subseteq A_S$.

Example. Hypergeometric expressions from *HypEx* are not closed under $\oplus$, but with respect to Hadamard-product $\odot$, etc. Hence we have,

$$(r \oplus_T s) \odot_T t = (r \odot_T t) \oplus_T (s \odot_T t) = (r \odot_X t) \oplus_T (s \odot_X t)$$

for all $r, s, t \in X$. The first equivalence follows from A_τ, the second from $A_{X \to T}$.

3 Symbolic summation as simplification

The embedding of admissible sequence-domains into the free term algebras T and S, as described in Sect. 2, serves as the basis for specifying the symbolic summation problem.

3.1 Indefinite summation

Now the problem of simplifying $\Sigma(f)$, where $f \in \mathbf{F}^{\mathbf{N}}$, can be described as finding a sum-quantifier free representation in the following sense.

a. Fix an admissible sequence-domain X that contains an expression t representing the summand sequence f, i.e., $v(t) = f$. In addition, let A be the set of axioms inducing equivalence in the sense described above.

b. Let $T = \mathit{Term}(X, \tau)$ and $S = \mathit{Term}(X, \sigma)$ be the free term algebras generated by X, and with axiom sets A_T and A_S as described above.

c. We say that t *is summable in* T *with respect to* A_S, if the following problem can be solved.

Indefinite symbolic summation problem
Given: $t \in X$,
find: $s \in T$ such that $s =_{A_S} \Sigma_S(t)$.

Evidently several variations can be formulated. For instance, one could relax the input condition to a given $t \in T$, or, in view of $\Sigma(t) = t * \mathbf{1} = \mathbf{1} * t$, more generally to convolution:

Convolution problem
Given: $t_1, t_2 \in T$,
find: $s \in T$ such that $s =_{A_S} t_1 * t_2$.

There are also several possibilities for solving the problem, depending on X and its signature a. For instance, it might be that $\Sigma \in a$ or $* \in a$. In this situation the summation problem finds a solution for *every* input $t \in X$.

Example. a. For all $k \in \mathbf{N}$ consider the summation $\sum_{j=0}^{k}(-1)^j\binom{a}{j}$, and choose for the sequence-domain $X := \mathit{GenFu}$. From the (ordinary) generating function relation $(1-x)^a = \sum_{j \geq 0}\binom{a}{j}(-1)^j x^j$ we have that the summand sequence is represented by $t := (1-x)^a$. The (ordinary) generating function representation of the constant sequence $\mathbf{1}$ is the geometric series $1/(1-x) = \sum_{j\geq 0} x^j$, thus one gets

$$\Sigma_S(t) = t *_S \mathbf{1} = (1-x)^a *_X \frac{1}{1-x} \in X\ . \tag{9}$$

b. If the given sum is slightly modified to $\sum_{j=0}^{k}(-1)^{k-j}\binom{a}{j}$, then in $X :=$ *GenFu* it can be represented as a convolution,

$$\frac{1}{1+x} *_S (1+x)^a = \frac{1}{1+x} *_X (1+x)^a\ .$$

c. For all $k \in \mathbf{N}$ consider the sum $\sum_{j=0}^{k} 1/(j+1)$, i.e., harmonic number summation, and choose for the sequence-domain $X :=$ *GenFu* as above. The summand sequence representation is delivered by the (ordinary) generating function relation $t := -1/x *_X \log(1-x) = \sum_{j\geq 0} x^j/(j+1)$, thus,

$$\Sigma_S(t) = t *_S \mathbf{1} = -\frac{1}{x} *_X \log(1-x) *_X \frac{1}{1-x} \in X\ . \tag{10}$$

Already at this elementary level the fundamental role of a *(canonical) simplifier* becomes evident. The importance not only lies in simplification, but also in preprocessing for conversion of the result into an equivalent expression from another domain.

Example. a. Applying a simplifier which takes care of the power rule, results in

$$(1-x)^a *_X \frac{1}{1-x} = (1-x)^{a-1} \in X\ .$$

Now it is an easy matter to convert the result into an elementary expression,

$$Convert((1-x)^{a-1}) = (-1)^k\binom{a-1}{k} \in EleEx\ .$$

It is also an easy matter to convert the binomial expression to an expression from *HypEx*:

$$Convert\left((-1)^k\binom{a-1}{k}\right) = \frac{(1-a)^{\bar{k}}}{(1)^{\bar{k}}} \in HypEx\ .$$

b. Analogously,

$$\frac{1}{1+x} *_X (1+x)^a = (1+x)^{a-1} \in X\ ,$$

which leads to

$$Convert((1+x)^{a-1}) = \binom{a-1}{k} \in EleEx\ .$$

c. In the harmonic number case it is not possible to obtain further substantial simplification to a rational function or to a power expression. Hence we present the result as a (Cauchy) product of a rational function and an elementary transcendental function (with a polynomial argument)

$$-\frac{1}{x} *_X \log(1-x) *_X \frac{1}{1-x} = -\frac{1}{x(1-x)} *_X \log(1-x)\ .$$

For instance, by using Gosper's algorithm one can *prove* that there exists no representation in form of a hypergeometric power series, i.e., as an element in *HypFu* (cf. Sect. 4). This can be rephrased as follows: In $X = GenFu$ there exists a solution to the harmonic number summation, which is not true if we take $X = HypEx$, the elementary hypergeometric expressions.

Remark. For algorithmic conversion between generating functions, recurrences and elementary expressions in the holonomic paradigm (Sect. 4), see the work of Salvy and Zimmerman (1994). An alternative approach is outlined by Koep (1992).

In situations where the signature a of X does not contain Σ or $*$, one standard approach to solve the summation problem is via telescoping.

Given: $t \in X$,
find: $s_0 \in T$ (or $s_0 \in X$) such that

$$t = \Delta(s_0)\ . \tag{11}$$

Because then $\Sigma(t) = \Sigma(\Delta(s_0)) = E(s_0) \oplus (-L(s_0)) =: s \in T$.

As an illustrating example we briefly discuss hypergeometric summation in difference field setting. The same example in the frame of Gosper's algorithm can be found in Sect. 4.

Example. The pair $(\mathbf{F}, \gamma)$ is called a *difference* field if $\gamma\colon \mathbf{F} \to \mathbf{F}$ is a field automorphism. Consider the following transcendental extension $(\mathbf{F}(x, y), \gamma)$ of $(\mathbf{F}, \gamma)$ where x and y are new symbols such that $\gamma c := c$ for all $c \in \mathbf{F}$, $\gamma x := x + 1$, and $\gamma y := r(x)y$ for some rational function $r \in \mathbf{F}(x)$. By its homomorphic extension, γ is a field automorphism of $\mathbf{F}(x, y)$. For notions, theoretical background, and algorithms concerning difference field extensions of this type, see the fundamental work of Karr (1981, 1985).

In view of the hypergeometric sequence defined by (7), consider the evaluation homomorphism *ev*: $\mathbf{F}(x, y) \to \mathbf{F}$ defined via homomorphic extension by $ev(c) := c$ for all $c \in \mathbf{F}$, $ev(x) := 0$, and $ev(y) := f_0$. This extends to *eval*: $\mathbf{N} \times \mathbf{F}(x, y) \to \mathbf{F}$ via

$$eval(n, R(x, y)) := R(x, y)[n] := ev(\gamma^n R(x, y)) \ .$$

For instance, y represents the sequence $(f_j)_{\geq 0}$ as defined in (7), because for all $j \in \mathbf{N}$:

$$y[j+1] = ev(\gamma^{j+1} y) = ev(r(x+j)(\gamma^j y)) = r(j)y[j] \ ,$$

and $y[0] = f_0$. Evidently $X := ((\mathbf{F}(x, y), \gamma), eval)$ is an a-algebra of type $(X, \{\oplus, \odot, L, E, \Delta\})$, where $\oplus$ and $\odot$ correspond to $+$ and $\cdot$, respectively, in the field $\mathbf{F}(x, y)$. For $t := R(x, y) \in X$ we have $L_X(t) := R(0, f_0)$, $E_X(t) := R(x+1, r(x)y)$, and $\Delta_X(t) := R(x+1, r(x)y) - R(x, y)$.

Now let us consider a specific instance, $\sum_{j=0}^{k}(-1)^j \binom{a+1}{j+1}$, a shifted variant of (4). For $f_j := (-1)^j \binom{a+1}{j+1}$ it is easily checked that $f_{j+1}/f_j = (j-a)/(j+2)$ and $f_0 = a + 1 \in \mathbf{F}$. Therefore, defining $r(x) := (x-a)/(x+2)$ the element $t := y \in \mathbf{F}(x, y)$ represents the summand sequence $(f_j)_{j \geq 0}$. Karr's algorithm solves $t = \Delta_x(s_0)$ for $s_0 = -(x+1)y/(a+1)$. Hence, $\sum_{j=0}^{k} f_j = eval(k, \sum_s(t)) = eval(k, E(s_0) \oplus_T (-L(s_0))) = E(s_0)[k] - L(s_0)[k] = -(k-a)/(a+1)(-1)^k \binom{a+1}{k+1} + 1$.

Not only from simple examples as this one it is evident that easy conversion between sequence-domains is desirable. More generally, there are many methods for dealing with symbolic summation (indefinite and definite) that make more or less explicit use of a strategy that implies domain changes in one or in several steps of the computation. Some illustrating examples are given in Sect. 5.

The best way to approach the problem of (easy) conversion is by using sequence-domains on which a *canonical simplifier* is defined. Consider, for example, the domain *HypEq*, where, as made more explicit in Sect. 4, the expressions are recurrence representations of minimal order. Another example concerns algorithms like Zeilberger's "fast" algorithm (Zeilberger 1990) that solve (definite) hypergeometric summation by delivering a linear recurrence with polynomial coefficients, i.e., a P-finite recursion that in most instances but not always is of order one. In order to decide whether there exists a hypergeometric evaluation, one additionally needs an algorithm that simplifies the recurrence to order one. Such an algorithm has been developed by Petkovšek (1992). The problem of establishing a canonical simplifier that reduces any P-finite recursion to minimal order still is not solved completely.

In a sense which is made precise below, one can consider the symbolic summation problem as the problem of extending canonical simplification to sum-expressions:

Indefinite symbolic summation (as canonical simplification)
Given: an admissible sequence-domain X with a canonical simplifier CS that extends to the term algebra $T = Term(X, \tau)$ with respect to the axiom set A_T.
Find: a subset $X_\Sigma \subseteq X$ together with an effective procedure SUM: $\Sigma_S(X_\Sigma) := \{\Sigma_S(t) \mid t \in X_\Sigma\} \to T$ such that

a. for all $t \in X_\Sigma$:

$$\Sigma_S(t) =_{A_s} SUM(t) , \tag{12}$$

b. X_Σ is maximal with respect to property (a).

This means, CS can be extended to all elements of $\Sigma_S(X_\Sigma)$ by defining

$$CS(\Sigma_S(t)) := CS(SUM(t))$$

for all $t \in X_\Sigma$.

Example. a. If $X := PolEx$ denotes the polynomial expressions from $HypEx$ then any canonical simplifier on X can be extended to all elements of $\Sigma_S(X)$, i.e., $PolEx$ is closed under summation.

b. If $X := HypEq$ then any canonical simplifier on X can be extended by $SUM =$ Gosper's algorithm to $\Sigma_S(X_\Sigma)$, where X_Σ can be defined as containing all recursions from X for which (11) is solvable.

Now we are in the position to give a precise definition of the notion "summation in closed form".

Indefinite summation in closed form
Let X, S, T be defined as in the description of the indefinite summation problem, and suppose that CS can be extended to $\Sigma_S(X_\Sigma)$ as above.

a. $t \in X$ is *summable in closed form with respect to SUM*, if $t \in X_\Sigma$, i.e., $\Sigma_S(t) =_{A_S} SUM(t) \in T$.

b. $CS(SUM(t))$ is said to be *the closed form of* $\Sigma_S(t)$ *with respect to SUM* ($t \in X_\Sigma$).

We want to remark that in various situations, where the canonical simplifier CS can be extended to $\Sigma_S(X_\Sigma)$ for $X_\Sigma \subseteq X$ with $X_\Sigma \neq X$ it might be possible to extend CS to the whole set $\Sigma_S(X)$, if the following problem can be solved algorithmically for any input.

Extracting the maximal summable part
Given: $t \in X \setminus X_\Sigma$,
find: $s \in X_\Sigma$ and $r \in X \setminus X_\Sigma$ such that

$$\Sigma_S(t) = SUM(s) \oplus_S \Sigma_S(r) \tag{13}$$

and where r satisfies some minimality criterion. In other words, the $s \in X_\Sigma$ can be extracted as a maximal summable part of $\Sigma_S(t)$.

In case it is possible to define a selector function on all representations of type (13), and if (r, s) is the selected pair, then

$$CS(SUM(s)) \oplus_S \Sigma_S(CS(r)) \in S$$

could serve as a canonical form for $\Sigma_S(t)$ in S.

Example. Take as sequence-domain X the rational function expressions from *HypEx*. Evidently X is an admissible a-algebra of type $(X, \{+, \odot, L, E, \Delta\})$, where we use the usual rational function $+$ for $\oplus$. Let t represent the summand sequence in (1), i.e., $t = (2j^2 + 3j + 1)/((j+1)^2(j+2)^2(j+3))$ where j is a free variable. Then it can be shown, see, e.g., Paule (1993), that

$$\Sigma_S(t) = \frac{6j^2 + 13j + 8}{2(j+1)(j+2)^2} + \Sigma_S\Big(\frac{3}{(j+3)^2}\Big) \tag{14}$$

is a minimal representation, where minimality is defined with respect to the degree of the denominator $(j+3)^2$. In Pirastu and Strehl (1996) an algorithm is presented which, given t with a generic numerator, singles out from representations of this type a particular one where the denominator degree of the rational part also is minimal, thus providing a canonical simplifier for the rational summation problem.

Note that evaluation at $k \in \mathbf{N}$ yields the result in the more familiar form,

$$\sum_{j=0}^{k} t[j] = \frac{6k^2 + 13k + 8}{2(k+1)(k+2)^2} + \sum_{j=0}^{k} \frac{3}{(j+3)^2} \ . \tag{15}$$

3.2 Definite summation

All what is said above applies to *definite* summation as well.

Example. Different from the alternating sign variant (4) the indefinite sum

$$\sum_{j=0}^{k} \binom{n}{j} \tag{16}$$

finds no closed form evaluation as a hypergeometric term, whereas it is an elementary fact that for the definite sum, where $n \in \mathbf{N}$, we have

$$\sum_{j=0}^{n} \binom{n}{j} = 2^n \ . \tag{17}$$

We need to extend the notion of sequence-domain, in order to be able to represent multi-indexed sequences. This extension is straightforward, we only give a brief sketch of the double-indexed case.

For representing $\mathbf{F}^{\mathbf{N}\times\mathbf{N}}$ we take a set of expressions X together with an evaluation function *eval* such that

$$eval\colon\ (\mathbf{N}\times\mathbf{N})\times X \to \mathbf{F},\quad ((n,k),t)\mapsto t[n,k]\ .$$

An element $f \in \mathbf{F}^{\mathbf{N}}$ can be considered a sequence from $\mathbf{F}^{\mathbf{N}\times\mathbf{N}}$ by defining for all $n, k \in \mathbf{N}$:

$$\bar{f}_{n,k} := f_n \quad (\text{or } \bar{f}_{n,k} := f_k)\ .$$

Therefore the one-indexed sequences naturally can be embedded in the double-indexed domain, or, more generally, all what we said about sequence-domains carries over to the multi-indexed case. This is also true with respect to operations, despite the fact that we now encounter a greater variety, this means two L-operators, a.s.o.

Let $(X, eval)$ be a sequence-domain for $\mathbf{F}^{\mathbf{N}\times\mathbf{N}}$. With respect to summation (16) we consider the unary operation Σ on X defined via the assignment function v as

$$v(\Sigma(t)) := \left(\sum_{j=0}^{k} t[n,j]\right)_{n,k\geq 0}$$

for $t \in X$. In view of the definite summation (17) another unary operator, *DefiniteSum*, has to be defined:

$$v(\mathit{DefiniteSum}(t)) := \left(\sum_{j=0}^{n} t[n,j]\right)_{n\geq 0}\ .$$

Therefore, the fundamental difference between the operators Σ and *DefiniteSum* is that the latter maps to that subdomain of X representing sequences of $\mathbf{F}^{\mathbf{N}}$.

Certainly, definite summation operators can be defined in several ways, e.g., replacing the upper summation bound n by some (linear) function in n. Nevertheless, the presented setting for indefinite summation is flexible enough to be carried over also to these variations. Instead of spelling out the details of the corresponding notions ("summable with respect to", "closed form", etc.) for multi-indexed sequences, in Sects. 4 and 5 we shall focus on presenting further examples that illustrate various features of symbolic summation that either have been only touched so far, or that add further evidence to the fundamental importance of conversion between different sequence-domains.

4 Holonomic generating functions

Hypergeometric sequences are special instances of P-finite sequences. In this section we have a closer look on various representations, especially generating function representation. For the sake of simplicity let us assume that any rational function from $\mathbf{F}(x)$ can be factored over $\mathbf{F}$. This means we can rewrite any expression from *HypEx* in the form (8) involving only a constant $\lambda \in \mathbf{F}$ instead of a rational function $rat(j)$.

For that situation the standard notation, e.g., Graham et al. (1994), for the corresponding (ordinary) generating function reads as

$$\lambda \, {}_mF_n \left(\begin{matrix} a_1, \ldots, a_m \\ b_1, \ldots, b_n \end{matrix} ; c\, x \right) := \sum_{j=0}^{\infty} \frac{a_1^{\bar{j}} \ldots a_m^{\bar{j}}}{b_1^{\bar{j}} \ldots b_n^{\bar{j}}} \lambda\, c^j x^j \,. \tag{18}$$

Dividing out the constant $\lambda \in \mathbf{F}$ would normalize the first entry of the underlying sequence to 1. Let *HypFu* denote the set of expressions of the left-hand side form. For $t \in HypFu$ and $k \in \mathbf{N}$ let $eval(k, t) := \langle x^k \rangle \lambda \, {}_mF_n$, i.e., the coefficient of x^k in $\lambda \, {}_mF_n$. Then $X := (HypFu, eval)$ is an a-algebra of type $(X, \{\odot, L, E, \Delta\})$, where the operations are defined according to those in *HypEx*.

We want to point out that *HypFu* is closed with respect to another unary operation we have not considered so far, namely differentiation $\frac{\mathrm{d}}{\mathrm{d}x}$ with respect to x:

$$\frac{\mathrm{d}}{\mathrm{d}x} {}_mF_n \left(\begin{matrix} a_1, \ldots, a_m \\ b_1, \ldots, b_n \end{matrix} ; c\, x \right) = c \frac{a_1 \ldots a_m}{b_1 \ldots b_n} \, {}_mF_n \left(\begin{matrix} a_1+1, \ldots, a_m+1 \\ b_1+1, \ldots, b_n+1 \end{matrix} ; c\, x \right) .$$

The elements of *HypFu* also satisfy so-called hypergeometric differential equations. These are equations of the type

$$P_d(x) y^{(d)} + \ldots + P_1(x) y' + P_0(x) y = 0 \,, \tag{19}$$

with polynomials $P_0, \ldots, P_d \in \mathbf{F}[x]$, not all 0. Therefore the hypergeometric generating functions, i.e., the sequence-domain *HypFu*, can be embedded in a much richer algebraic structure, the ring of D-finite power series. A power series $y = \sum_{j=0}^{\infty} f_j x^j \in \mathbf{F}[[x]]$ is said to be *D-finite* (or *holonomic*) if it satisfies a differential equation of type (19), see Stanley (1980). D-finite power series include algebraic and elementary transcendental functions and satisfy nice closure properties. For instance, they are closed under $\oplus$, $\odot$, $*$, differentiation, and integration. Therefore an efficient methodology for the algorithmic treatment of generating functions is to work with D-finite power series (finitely) presented by their differential equation (19) together with its initial values. Such a tool is provided in form of the Maple package "gfun". See Salvy and Zimmermann (1994) for a survey and nontrivial examples.

Thus let *GenFu* be the set of tuples of coefficients encoding the differential equation (19) together with the initial values.

Then we have that $X := (GenFu, eval)$, *eval* as above, is an a-algebra of type (X, σ), σ the standard signature. We also want to note that, in general, if $y = \sum_{j=0}^{\infty} f_j x^j \in GenFu$ then $L_X(y) := y[0]/(1-x)$, $E_X(y) := 1/x *_X (y - y[0])$, and $\Delta_X(y) = 1/x *_X ((1-x) *_X y - y[0])$ (recall that $y[0] = f_0$).

A formal power series $y = \sum_{j=0}^{\infty} f_j x^j \in \mathbf{F}[[x]]$ is D-finite if and only if $(f_j)_{j \geq 0}$ is P-finite (e.g., Stanley 1980). *P-finite* means that there exist finitely many polynomials $p_0, \ldots, p_d \in \mathbf{F}[x]$, with $p_d \neq 0$, such that for all $j \in \mathbf{N}$:

$$p_d(j) f_{j+d} = p_{d-1}(j) f_{j+d-1} + \ldots + p_0(j) f_j \ . \tag{20}$$

The hypergeometric sequences correspond to the first nontrivial order $d = 1$, i.e., in view of (7) we have that $r(x) = p_0(x)/p_1(x)$. Hence, we define *HypEq* to consist of expressions of the form

$$(p_1(j)E - p_0(j)I, 0; c) \ , \tag{21}$$

where $c \in \mathbf{F}$ stands for the initial value $f_0 = c$. For $t \in HypEq$ the evaluation function *eval* is defined by computing $t[n]$ iteratively from the corresponding recurrence of order 1. It is easy to see that $X := (HypEq, eval)$ forms an a-algebra of type $(X, \{\odot, L, E, \Delta\})$. For $t = (p_1(j)E - p_0(j)I, 0; c) \in HypEq$ and $s = (q_1(j)E - q_0(j)I, 0; d) \in HypEq$ we have

$$t \odot_X s := (p_1(j)q_1(j)E - p_0(j)q_0(j)I, 0; c\ d) \ ,$$
$$E_X(t) := (p_1(j+1)E - p_0(j+1)I, 0; c\ p_0(0)/p_1(0)) \ ,$$

and

$$\Delta_X(t) := (p_1(j+1)(p_0(j) - p_1(j))E - p_0(j)(p_0(j+1) - p_1(j+1))I, \\ 0; c\,(p_0(0) - p_1(0))/p_1(0)). \tag{22}$$

We conclude this section by considering hypergeometric summation again.

Example. Given a hypergeometric sequence $f := (f_j)_{j \geq 0} \in \mathbf{F}^{\mathbf{N}}$, the hypergeometric summation problem usually is described as the problem of finding a hypergeometric solution $g := (g_j)_{j \geq 0} \in \mathbf{F}^{\mathbf{N}}$ of $f = \Delta(g)$. Then, by telescoping, $\sum_{j=0}^{k} f_j = g_{k+1} - g_0$ for all $k \in \mathbf{N}$. Gosper's algorithm (Gosper 1978, Graham et al. 1994) finds the solution g in case it exists.

Taking as the representing sequence domain $X := HypEq$ the situation is as follows. Let $t := (p_1(j)E - p_0(j)I, 0; c) \in HypEq$ represent f as in (21). According to the general strategy described above, we try to find an $s_0 \in X$ such that $t = \Delta_X(s_0)$. Take as an "Ansatz" for $s_0 := (q_1(j)E - q_0(j)I, 0; d) \in HypEq$ with $d \in \mathbf{F}$ and with polynomials $q_0, q_1 \in \mathbf{F}[x]$ with undetermined coefficients. Then, by (22), we have to determine q_0, q_1, d such that

$$\frac{p_0(j)}{p_1(j)} = \frac{q_0(j)(q_0(j+1) - q_1(j+1))}{q_1(j+1)(q_0(j) - q_1(j))} \tag{23}$$

and

$$c = \frac{d\,(q_0(0) - q_1(0))}{q_1(0)} \ . \tag{24}$$

Paule and Strehl (1995) explain how one gets q_0, q_1 by applying the Gosper–

Petkovšek representation of a rational function to (23); d easily is computed from (24). There are several methods to show that (23) can be solved for polynomials q_0, q_1 if and only if a solution $s_0 \in X$ of $t = \Delta_X(s_0)$ exists. One approach, see Paule (1993) or Paule and Strehl (1995), puts a special emphasis on canonical forms by embedding the problem in the more general context of *greatest-factorial factorization* (GFF) of polynomials, a shift operator analogue of square-free factorization. We conclude this example by considering again, $\sum_{j=0}^{k}(-1)^j\binom{a+1}{j+1}$, the example from Sect. 3. Here $t := ((j+2)E - (j-a)I,$ $0; a+1)$, and one can check via (22) that $s_0 = ((j+1)E - (j-a)I, 0; -1)$ is the solution, which easily can be converted into the binomial expression

$$g_j := (-1)^{j+1}\binom{a}{j}.$$

Remark. The approach of viewing hypergeometric summation as canonical simplification can be extended also for the treatment of so-called q-analogues of binomial or hypergeometric summations; see Paule and Strehl (1995). For instance, in additive number theory summations involving Gaussian polynomials, so-called q-binomial coefficients, play an important role. For a comprehensive survey, see Andrews (1986); for a recent algorithmic application, Paule (1994). A different q-approach, following Gosper's original presentation and ideas of Zeilberger, can be found in Koornwinder (1993). Indefinite q-hypergeometric summation also can be treated within the theory of Karr (1981).

5 Examples

This section provides illustrating examples of strategies that make flexible use of conversion between appropriate sequence domains.

5.1 On a conjecture of Knuth

Consider the difference $D_n := R_n - L_n$ of two finite hypergeometric sums, where

$$L_n := \sum_{k=0}^{n}\binom{2k}{k}$$

and

$$R_n := \frac{4}{3}(-4)^n \sum_{k=0}^{\lfloor(3n+2)/4\rfloor}\binom{k-1/2}{n}\left(-\frac{1}{3}\right)^k.$$

From a generating function observation, Knuth (pers. comm. 1994) conjectured that for all $n \in \mathbf{N}$,

$$0 \leq D_{n+4} < D_n. \tag{25}$$

Because of the fact that all the initial values D_0, D_1, D_2, D_3 are less than 1, the conjecture immediately implies the curious identity $L_n = \lfloor R_n \rfloor$.

It turns out that the monotonicity part of (25) finds a surprisingly simple, almost mechanical proof, namely by deriving with a computer program the inhomogeneous recurrence representations of both sums.

We want to remark that based on these recurrence representations also the positivity part, $0 \leq D_n$, of (25) can be settled, but this time using nontrivial classic hypergeometric series machinery. Hence, for the complete proof of (25) we refer the interested reader to Paule (1996).

We prove $D_{n+4} < D_n$ for the case $n = 4m$ ($n \in \mathbf{N}$), the remaining cases work analogously. In order to avoid factorials involving rational numbers, it is convenient to rewrite R_n as

$$R_n = \frac{4}{3}\binom{2n}{n} \sum_{k=0}^{\lfloor (3n+2)/4 \rfloor} \left(\frac{1}{3}\right)^k \binom{n}{k} \binom{2n}{2k}^{-1} .$$

This equality easily can be checked by verifying equality between the k-th summands.

For $m \in \mathbf{N}$ let $l_m := L_{4m}$, $r_m := R_{4m}$, and $d_m := D_{4m}$. Applying the package from Paule and Schorn (1994) one computes that

$$l_{m+1} - l_m = a(m) \tag{26}$$

and

$$r_{m+1} - r_m = a(m) - b(m) , \tag{27}$$

where

$$a(m) := 16(147 + 784m + 1302m^2 + 680m^3)\frac{(8m+1)!}{(4m)!(4+4m)!}$$

and

$$b(m) := \frac{8}{27}(8m+7)\left(\frac{1}{3}\right)^{3m} \frac{(2m+1)!(2+6m)!}{m!(1+3m)!(4+4m)!} .$$

Because of

$$d_m - d_{m+1} = b(m) ,$$

which is obtained by combining (26) and (27), the proposition follows immediately from $b(m) > 0$.

5.2 Interlacing sequences

With this example we want to illustrate that for computing a closed form for a sum it might be helpful to consider, besides standard operations like termwise addition etc., another natural sequence operation, namely *interlacing*. By interlacing the sequences $(s_k^{(0)})_{k\geq 0}$ and $(s_k^{(1)})_{k\geq 0}$ we mean $(s^{(k \bmod 2)})_{k\geq 0}$.

Let us consider the following encoding of the Fibonacci numbers which originally is due to I. Schur (see also Andrews 1986),

$$\sum_k (-1)^k \binom{n}{\lfloor (n-5k)/2 \rfloor} . \tag{28}$$

In view of the simple *RecEq* representation for the Fibonacci numbers, it is our goal to derive automatically a *RecEq* representation for (28). Because Zeilberger's algorithm for finding hypergeometric sums does not apply directly, we first have to eliminate the floor function. This is done by splitting the sum according to even and odd n and k. Then by applying the algorithm from Paule and Schorn (1994) we get

$$(E^3 - 7E^2 + 13E - 4I, 0; (1, 2, 6)) = \sum_{k'} (-1)^{2k'} \binom{2n'}{n' - 5k'} , \tag{29}$$

$$(E^3 - 7E^2 + 13E - 4I, 0; (0, 0, -1)) = \sum_{k'} (-1)^{2k'+1} \binom{2n'}{n' - 5k' - 2} , \tag{30}$$

$$(E^3 - 7E^2 + 13E - 4I, 0; (1, 3, 10)) = \sum_{k'} (-1)^{2k'} \binom{2n'+1}{n' - 5k'} , \tag{31}$$

$$(E^3 - 7E^2 + 13E - 4I, 0; (0, 0, -2)) = \sum_{k'} (-1)^{2k'+1} \binom{2n'+1}{n' - 5k' - 2} . \tag{32}$$

For combining that to the desired *RecEq* representation for (28) we use the package by Nemes and Petkovšek (1996) to compute

$$\begin{aligned} \mathit{Interlace}((&E^3 - 7E^2 + 13E - 4I, 0; (1, 2, 6)) \oplus \\ &(E^3 - 7E^2 + 13E - 4I, 0; (0, 0, -1)), \\ &(E^3 - 7E^2 + 13E - 4I, 0; (1, 3, 10)) \oplus \\ &(E^3 - 7E^2 + 13E - 4I, 0; (0, 0, -2))) \end{aligned}$$

and find

$$(E^2 - E - I, 0; (1, 1)) ,$$

the standard *RecEq* representation of the Fibonacci numbers.

5.3 Recursion to elementary expression

We consider problem E 3065 from the *American Mathematical Monthly* (vol. 90, pp. 649).

Let $n \geq 0$ and $a \geq n + 1$. Find a closed formula for

$$\sum_{k=0}^{n} (-1)^k \binom{a}{k} \binom{a-k-1}{n-k} / (k+1) . \tag{33}$$

For alternative solutions see *American Mathematical Monthly* (vol. 93, pp. 387) and also Roy (1988) where this and other problems serve to show how the theory of generalized hypergeometric series can be used to find closed forms for binomial sums. Here we solve the problem in a different way, first by transforming the sum to *RecEq* representation, then solving the derived recurrence equation thus illustrating a conversion from a *RecEq* representation to an *EleEx* one.

Let $s(n)$ stand for the sum (33). Then by applying the package by Paule and Schorn (1994) we get

$$s(n) + s(n+1) = \binom{a}{n+1}/(n+2) . \tag{34}$$

Now we are faced with the problem to transform $(E + I, \binom{a}{n+1}/(n+2); 1)$ to an elementary expression. The success of this step depends on whether *EleEx* contains $s(n)$. Fortunately, (34) can be solved by elementary methods. The solution of the corresponding homogeneous equation $s(n) + s(n+1) = 0$ is $(-1)^n c$ with a constant c. To get a particular solution of (34) it is enough to consider the relation

$$\binom{a}{n+1} + \binom{a}{n+2} = \binom{a+1}{n+2} = \frac{a+1}{n+2}\binom{a}{n+1} ,$$

which shows that $\binom{a}{n+1}/(a+1)$ is a solution of (34). Finally, observing $s(0) = 1$ one gets

$$s(n) = \frac{1}{a+1}\left(\binom{a}{n+1} + (-1)^n\right) .$$

We remark that there are recent results that provide algorithmic tools for the transformation *RecEq* $\to$ *EleEx* (Abramov 1989, Petkovšek 1992).

5.4 Difference field extensions

In order to illustrate the most general available algorithm for solving the indefinite summation problem we choose exercise 6.53 from Graham et al. (1994):

Find a closed form for

$$\sum_{j=0}^{m} \binom{n}{j}^{-1} (-1)^j h_{j-1} , \tag{35}$$

when $0 \le m \le n$, and where $(h_j)_{j \ge -1}$ is the harmonic number sequence with $h_{-1} = 0$.

The algorithm we use was developed by Karr (1981). A first implementation was written in Reduce by Gärtner (1986). A Mathematica implementation by E. Eichhorn is in preparation. The algorithm starts out with $(\mathbf{F}, \gamma) = (\mathbf{Q}(n)(j), E)$ which is considered as a difference field with the shift operator E acting on j.

Analyzing the input this difference field is extended by the solutions of the recurrence equations

$$\left(\frac{n-j}{j+1}E - I, 1; 0\right) \quad \text{and} \quad (j+1)E - (j+1)I, 1; 1) \ ,$$

corresponding to the hypergeometric and harmonic number part of the summand. In this extension the algorithm finds $s(j)$ such that

$$s(j) - s(j-1) = \binom{n}{j}^{-1} (-1)^j h_{j-1} \ .$$

By telescoping for (35) one gets

$$s(m) - s(-1) = \frac{n+1}{(n+2)^2} \left(\frac{(-1)^m}{\binom{n+1}{m+1}}((n+2)h_m - 1) - 1 \right) \ .$$

5.5 Computing with D-finite representations

Prodinger (1994) surveyed the history and the method of closed form evaluation of

$$s(n) := \sum_{j\geq 0} \binom{n}{j}\binom{2j}{j}(-1)^j/2^j \ . \tag{36}$$

It is very instructive to read Prodinger (1994) to get an impression of the variety of possible approaches for this particular problem. By using the program package "gfun" by Salvy and Zimmermann (1994) and following the pattern of the example using Euler's transform in Flajolet and Salvy (1996), we give a modified version of the solution presented in Prodinger (1994: sect. 4). Euler's transform of a generating function $u(z)$ reads as

$$\langle z^n \rangle \frac{1}{1-z} u\left(\frac{z}{z-1}\right) = \sum_{k=0}^{n} \binom{n}{k} (-1)^k \langle z^k \rangle u(z) \ . \tag{37}$$

Let $y(z)$ be the ordinary generating function of $u = \left(\binom{2j}{j}/2^j\right)_{j\geq 0}$. Notice that the ordinary generating function of $s(n)$ is the Euler transform of $y(z)$. The following five steps with "gfun" show how to derive from the recurrence representation of u a recurrence representation for s.

A *RecEq* representation, i.e., the hypergeometric recurrence for u is derived easily:

```
> rec:= (k+1)*u(k+1) - (2*k+1)*u(k)=0;
         rec := (k + 1) u(k + 1) - (2 k + 1) u(k) = 0 .
```

The function `rectodiffeq` converts this representation to a differential equation representation of the corresponding generating function $y(z)$:

```
> rectodiffeq(rec,u(k),y(z));
              y(z) + (2 z - 1) D(y)(z).
```

Perform the substitution $z \mapsto z/(z-1)$:

```
> algebraicsubs(",y*(z-1)-z,y(z));
                      2
              -y(z) + (z  - 1) D(y)(z).
```

Derive a differential equation representation for $y(z)$ multiplied by $1/(1-z)$:

```
> `diffeq*diffeq`(",(1-z)*y(z)-1,y(z));
                          2
              -y(z) z + (-z  + 1) D(y)(z).
```

Finally we convert this differential equation representation into a recurrence representation:

```
> diffeqtorec(",y(z),s(n));
              {(-n - 1) s(n) + (n + 2) s(n + 2), s(1) = 0}.
```

From this *RecEq* representation one gets the *EleEx* representation:

$$s(n) = \begin{cases} \frac{1}{2^n}\binom{n}{n/2} & \text{if } n \text{ even,} \\ 0 & \text{if } n \text{ odd.} \end{cases}$$

5.6 Completely computerized evaluation

A completely computerized, alternative evaluation of (36) is done as follows. Compute sufficiently many initial values $S(n)$, such that the recurrence representation can be guessed by "gfun".

```
> listtorec([1,0,1/2,0,3/8,0,5/16,0,35/128],u(n));
  [{u(0) = 1, u(1) = 0, (-n - 1) u(n) + (n + 2) u(n + 2)}, ogf]
```

According to Yen (1993), to conclude

$$s(n) = u(n) \tag{38}$$

for all n, it is enough to verify it up to a sufficiently large n. Computing this bound according to Yen (1993), and verifying (38) for n less than this bound completes the verification of the guessed result.

Acknowledgement

P. Paule is supported in part by grant P7220 of the Austrian Fonds zur Förderung der Wissenschaftlichen Forschung.

References

Abramov, S. A. (1989): Rational solutions of linear difference and differential equations with polynomial coefficients. USSR Comput. Math. Math. Phys. 29: 7–12.

Andrews, G. E. (1986): q-Series: their development and application in analysis, number theory, combinatorics, physics, and computer algebra. American Mathematical Society, Providence (CBMS regional conference series, vol. 66).

Buchberger, B., Loos, R. (1983): Algebraic simplification. In: Buchberger, B., Collins, G., Loos, R. (eds.): Computer algebra, symbolic and algebraic computation, 2nd edn. Springer, Wien New York, pp. 11–43.

Egorychev, G. P. (1984): Integral representation and the computation of combinatorial sums. American Mathematical Society, Providence (Translations of mathematical monographs, vol. 59).

Flajolet, P., Salvy, B. (1996): Computer algebra libraries for combinatorial structures. J. Symb. Comput. 20: 653–672.

Gärtner, J. (1986): Summation in finite terms – presentation and implementation of M. Karr's algorithm. Diploma thesis, Johannes Kepler University, Linz, Austria.

Gosper, R. W. Jr. (1978): Decision procedure for indefinite hypergeometric summation. Proc. Natl. Acad. Sci. U.S.A. 75: 40–42.

Gould, H. W. (1972): Combinatorial identities, a standardized set of tables listing 500 binomial coefficient summations, revised edn. West Virginia University, Morgantown.

Graham, R. L., Knuth, D. E., Patashnik, O. (1994): Concrete mathematics: a foundation for computer science, 2nd edn. Addison-Wesley, Reading, MA.

Karr, M. (1981): Summation in finite terms. J. ACM. 28: 305–350.

Karr, M. (1985): Theory of summation in finite terms. J. Symb. Comput. 1: 303–315.

Knuth, D. E. (1973): The art of computer programming, vol. 1, fundamental algorithms, 2nd edn. Addison-Wesley, Reading, MA.

Koepf, W. (1992): Power series in computer algebra. J. Symb. Comput. 13: 581–603.

Koornwinder, T. H. (1993): On Zeilberger's algorithm and its q-analogue. J. Comput. Appl. Math. 48: 91–111.

Lafon, J. C. (1983): Summation in finite terms. In: Buchberger, B., Collins, G., Loos, R. (eds.): Computer algebra, symbolic and algebraic computation, 2nd edn. Springer, Wien New York, pp. 71–77.

Nemes, I., Petkovšek, M. (1996): RComp: a Mathematica package for computing with recursive sequences. J. Symb. Comput. 20: 745–754.

Paule, P. (1993): Greatest factorial factorization and symbolic summation I. Tech. Rep. 93-02, RISC Linz, Johannes Kepler University, Linz, Austria.

Paule, P. (1994): Short and easy computer proofs of the Rogers–Ramanujan identities and of identities of similar type. Electron. J. Combinatorics 1: R10.

Paule, P. (1996): Proof of a conjecture of Knuth. J. Exp. Math. 5: 83–89.

Paule, P., Schorn, M. (1996): A Mathematica version of Zeilberger's algorithm for proving binomial coefficient identities. J. Symb. Comput. 20: 673–698.

Paule, P., Strehl, V. (1995): Symbolic summation – some recent developments. In: Fleischer J., Grabmeier, J., Hehl, F., Kuechlin, W. (eds.): Computer algebra in science and engineering: algorithms, systems, and applications. World Scientific, Singapore, pp. 138–162.

Petkovšek, M. (1992): Hypergeometric solution of linear recurrences with polynomial coefficients. J. Symb. Comput. 14: 243–264.

Pirastu, R., Strehl, V. (1996): Rational summation and Gosper–Petkovšek representation. J. Symb. Comput. 20: 617–636.

Prodinger, H. (1994): Knuth's old sum – survey. EATCS Bull. 54: 232–245.

Roy, R. (1988): Binomial identities and hypergeometric series. Am. Math. Monthly 94: 36–46.

Salvy, B., Zimmermann, P. (1994): GFUN: a Maple package for the manipulation of generating and holonomic functions in one variable. ACM Trans. Math. Software 20: 163–177.

Schorn, M. (1995): Contributions to symbolic summation. Diploma thesis, Johannes Kepler University, Linz, Austria.

Slater, L. J. (1966): Generalized hypergeometric series. Cambridge University Press, Cambridge.

Stanley, R. P. (1980): Differentiably finite power series. Eur. J. Comb. 1: 175–188.

Wilf, H. S. (1990): Generating functionology. Academic Press, San Diego.

Yen, L. (1993): Contributions to the proof theory of hypergeometric identities. Ph.D. thesis, University of Pennsylvania, Philadelphia, PA.

Zeilberger, D. (1990): A fast algorithm for proving terminating hypergeometric identities. Discrete Math. 32: 321–368.

Indexes in sums and series: from formal definition to object-oriented implementation

O. Caprotti

1 Introduction

Applicative mathematics is done with indexes. Physics, chemistry, and engineering sciences yield a wide range of applications where expressions containing indexes are used for formalizing and computing. Nevertheless, the available computer algebra systems often show little skill in dealing with expressions that contain indexes. Typical problems include: incorrect notion of variable binding and lack of knowledge about the domains over which the variables range. Wang (1990) and Cioni and Miola (1990) pointed out these deficiencies. In this paper, we show how the object-oriented design methodology proposed by the TASSO project can be applied successfully to this case.

Computer algebra systems often do not recognize bindings so that, for instance, summation of an indexed variable is turned into a sum of a constant:

$$\text{assign } x_k \text{ to } y\ ,$$

$$\sum_{k=0}^{10} y \rightarrow 11 x_k\ .$$

A symbolic environment should treat y as an indexed indeterminate having name x and index k. This is easily accomplished by an object-oriented approach in which y is bound to an object representing indexed indeterminates. In such a way, when performing a summation of an indexed indeterminate, a method that checks both the index of the sum and that of the summand is invoked and produces the right result.

One of the most attractive features of the object-oriented view for symbolic manipulation, is the dispatching of methods that is done at run-time according to the different possible arguments. Such a feature is extremely useful in order to help the system when evaluation is ambiguous. Consider

$$\sum_{k=-1}^{10} \int x^k \, \mathrm{d}x \rightarrow \text{division by zero}\ .$$

Since the system evaluated the integral first, it ended up with a problem caused by the index range of the summation. Instead, if it would have swapped summation with integration – knowing that linear operators distribute over a finite sum – it would have produced the correct answer. By run-time dispatching, the method performing the summation checks whether the summand contains an operator that distributes over the sum. If the sum is found to be ambiguous, then it attempts to re-order the evaluation and to solve it. The Common Lisp Object System (CLOS) (Bobrow et al. 1988), the object-oriented extension of Lisp chosen for the implementation, provides a run-time dispatching mechanism.

Furthermore, object-oriented languages emphasize type information. In the symbolic manipulation context, this helps to ensure correctness by preventing erroneous operations. For example, the method invoked to perform differentiation of an indexed expression with respect to an indexed identifier, like

$$\frac{\partial}{\partial x_h}\left(\sum_{k=0}^{10} x_k\right),$$

should output different answers depending on the index. The same method could also guard that differentiation is not done with respect to the index, assumed to be an integer and therefore not continuous. Currently, one can expect the following behavior:

$$\frac{\partial}{\partial x_h}\left(\sum_{k=0}^{10} x_k\right) \to 0 .$$

Instead, it would be desirable, from a mathematical point of view, to have a method that outputs either the answer 1, in case $0 \leq h \leq 10$, or the value 0 otherwise.

Attempts to solve these problems resulted in special-purpose toolkits built on top of the various computer algebra systems: Miola (1983) considers the case of handling indefinite summations with indexed variables, and Wang (1991) and Shen et al. (1992) develop a special package tailored for neural networks applications.

In our experience, most cases of misbehavior can be related to insufficient type checking of the bound variables appearing in the expression: the decision to apply a procedure does not depend on the actual type of the treated object. The problems that the systems encountered in handling summations could have been avoided if, during the design phase of data types and procedures, the underlying mathematical definitions of indexed indeterminate, of indexed functions would have been used and properties of summation taken into account. This is where the design methodology outlined by TASSO can be effectively applied because it advocates the integration of the three development phases of formal definition, specification, and implementation of the abstract data types modelling the axiomatization. The object-oriented approach to software development naturally supports this integration.

In object-oriented design (Limongelli and Temperini 1992), the primary issue is the specification of the classes of objects that will be manipulated by the system. This step is characterized by the abstract level typical of a formal, mathematical description. In case of mathematical objects, the specification of the abstract data types follows directly from the mathematical formalization and can be implemented in an object-oriented environment through *classes*.

In this fashion, one achieves the combination of modular and typing aspects because:

- *Inheritance*, the principle by which a class may be defined as an extension or a restriction of another, can be fruitfully used when the classes are structured in a hierarchical way. Algebraic domains, like those in the category hierarchy used in Scratchpad (Jenks and Trager 1981), are an example of such a situation.
- *Polymorphism*, the mechanism that allows operations on objects belonging to more than one class, can be exploited by providing functions defined abstractly over the proper algebraic domain.

The paper is structured according to the phases in the object-oriented design methodology. Section 2 provides the formal characterization of indexed variables, indexed functions and summation. Then, the specification of the abstract data types is derived and presented in Sect. 3. Hence, Sect. 4 shows the CLOS implementation of classes and methods, naturally derived from the specification. Section 5 is devoted to concluding remarks and perspectives for future work.

2 A formal characterization of indexed symbolic objects, sums, and series

In this section, we define the main mathematical objects that the symbolic system deals with: indexed variables, indexed functions and summations. In this phase, the definitions are given at the most abstract level so that the specification of abstract data types follows almost directly. We also recall the basic rules to manipulate sums and series (Miola 1983).

2.1 Indexed objects

When mathematicians use indexes they make implicit assumptions which ensure that a notation like x_k makes sense in a computation. We try to formalize the intuition behind the usage of indexed objects. It is clear that from the purely syntactical point of view, no restriction can be imposed on indexed notations. For instance, they can be used to represent sets or sequences that are related to some real or rational number, like: $S_{\sqrt{2}} = \{i \leq \sqrt{2} \mid i \in \mathbf{R}\}$.

However, in usual computations, an indexed variable refers to a correspondence between the enumerable set, that is the range of the index, and the domain of the variable. Therefore, the range of indexes is assumed to be a subset of the integers. No computational meaning is assigned to a notation like $S_{\sqrt{2}}$. On the other hand, it is possible to build complex indexes by using arithmetical expressions as in x_{k+2*h}, where h and k are both indexes.

The following characterization of indexed object formalizes all the considerations above. From an abstract point of view, an indexed object can be considered an ordered pair formed by a symbol and an index, which is a term ranging on the integers: such will be the implementation of indexed variables through the abstract data type (ADT) **Indexed_Var**.

Consider a finite set $\mathcal{D}_0 = \{\mathbb{N}, \mathbb{Z}, \mathbb{Q}, \mathbb{R}, \mathbb{C}, \ldots\}$ of primitive types, and let $\mathcal{D}$ be the smallest set such that: $\mathcal{D}_0 \subseteq \mathcal{D}$, and for all $\alpha, \beta \in \mathcal{D}$, the type $(\alpha \to \beta) \in \mathcal{D}$.

For each type d in $\mathcal{D}$, let $\mathcal{V}_d$ be an enumerable set of variables of type d and similarly, let $\mathcal{C}_d$ be an enumerable set of constants of type d such that $\mathcal{V}_d \cap \mathcal{C}_d = \emptyset$. We can define the set, $\mathcal{T}_d$, of *terms of type d* as the smallest set such that:

- $\mathcal{V}_d \cup \mathcal{C}_d \subseteq \mathcal{T}_d$,
- if $d = (\mathbb{Z} \to s)$, then the set $\mathcal{IV}_d \subseteq \mathcal{T}_d$, where $\mathcal{IV}_d$ is the set of pairs $\langle v, k\rangle$, such that $v \in \mathcal{V}_s$ and $k \in \mathcal{T}_{\mathbb{Z}}$,
- for every term $f \in \mathcal{T}_{(c \to d)}$ and every $t \in \mathcal{T}_c$, the application $f(t) \in \mathcal{T}_d$.

Elements in $\mathcal{IV}_d$ are called *indexed variables* if the variable in the pair $\langle v, k\rangle$ is of a basic type, namely $d \in \bigcup_{s \in \mathcal{D}_0} \mathcal{T}_{(\mathbb{Z} \to s)}$; otherwise, they are called *indexed functions*.

Since in a safe computing environment, unpleasant situations like x_x or $y_x(y)$ should be avoided, we impose the restriction that the sets of terms of different types are pairwise disjoint.

2.2 Sums and series

We now turn to the definition of the repeated sum, normally denoted by the Sigma notation, as:

$$\sum_{k:\,\Delta} A$$

where k is a term in $\mathcal{T}_{\mathbb{Z}}$, Δ is a relation over an ordered set (namely, the integers) and $k{:}\ \Delta$ is the set of all k's satisfying Δ, A is a term of type $\mathcal{D}_{(\mathbb{Z} \to r)}$ (for r a ring). Our attention will be restricted to sums of *ring elements*. In a sum, A is called the *summand*, and commonly, is an expression containing k that is called the *index* of the sum.

The index k of a summand can be considered *bound* by the sigma sign. This accounts for calling Sigma a quantifier, and the index a *dummy* variable. As with all bound variables, the index is unrelated to other occurrences outside of the Sigma-notation and can be renamed.

A sum is completely specified through the index, the relation over the index and the summand. We must distinguish among:

- *finite sum*, when the relation $k{:}\ \Delta$ is true only for k ranging in a finite set

$\{k_1, \ldots, k_n\}$. The Sigma-notation represents the iterated sum:

$$\sum_{k:\,\Delta} A = A(k_1) + \ldots + A(k_n) .$$

Note that, when k: Δ reduces to the empty set, the sum is defined to be equal to zero. Typical examples of this kind of relations are those describing intervals as: $0 \leq j \leq 11$.

– *Infinite series*, when the relation k: Δ is true for infinite values of k, say $\{\ldots, k_{-1}, k_0, k_1, \ldots\}$. The Sigma-notation represents a series whose meaning is related to the behavior of the limit of the succession of its partial sums. When this limit does not exist, the infinite sum does not exist and the series is called a "divergent" series.

Straightforwardly from this, one can derive the definition of the ADT **Sigma** of sum objects to be manipulated.

2.3 Laws of manipulation of sums and series

Rather than the conventional evaluation of sums and series, as it is done in most computer algebra systems, our aim is to define a set of rules to transform expressions involving sums by manipulating either the summand or the relation over the index. Manipulation rules are in general valid for finite sums and in some cases, for series that are convergent. Table 1 lists the three basic rules for the manipulation of Sigma expressions.

The rule **Interchange** is a very general rule that allows to distribute a linear operator ∇ over a sum. It is valid for convergent series.

The rule of associativity is a special case of the **Interchange** rule, when the operator is addition:

$$\sum_{k:\,\Delta} (A + B) = \sum_{k:\,\Delta} A + \sum_{k:\,\Delta} B .$$

The general form of the associativity rule is the rule for interchanging the order of summations when the indexes do not depend on each other:

$$\sum_{k:\,\Delta} \Big(\sum_{h:\,\Theta} A \Big) = \sum_{h:\,\Theta} \Big(\sum_{h:\,\Theta} A \Big) .$$

Abstractness requires the specification of the rule **Interchange** in the most gen-

Table 1. Rules for Sigma manipulation

Interchange	$\sum_{k:\,\Delta} \nabla A = \nabla \sum_{k:\,\Delta} A,$	∇ is a linear operator
Distribute	$\sum_{k:\,\Delta} c \cdot A = c \cdot \sum_{k:\,\Delta} A,$	c is a constant w.r.t. k
Split Domains	$\sum_{k:\,\Delta} A + \sum_{k:\,\Theta} A = \sum_{k:\,\Delta \vee \Theta} A + \sum_{k:\,\Delta \wedge \Theta} A.$	

eral form. Subsequently, the implementation will produce different methods according to the different special cases that depend on the linear operator. For example, when ∇ is addition, the method takes associativity into account, while if ∇ is a linear operator such as integration, then the method implements the general rule.

Rule **Distribute** states that, whenever the summand is multiplied by a constant with respect to the index, then this constant can be directly multiplied by the sum. Namely, this rule allows to move constants in and out of the sigma quantifier. It is valid for arbitrary series. One specialization is the distributive law for product of sums which can also be read as a law of factorization for products of sums:

$$\sum_{h:\Theta} B \cdot \sum_{k:\Delta} A = \sum_{k:\Delta}\left(A \cdot \sum_{h:\Theta} B\right) = \sum_{k:\Delta}\sum_{h:\Theta}(A \cdot B) .$$

The rule **Split Domains**, that manipulates the domain of the sum, can be applied for dividing the domains into subdomains and, conversely, for combining several domains into a single one. De facto, in most applications the condition $k\colon \Delta \wedge \Theta$ is equivalent to the empty set, thus this summand disappears. No restriction is imposed for the application of this rule which is valid for arbitrary series.

The manipulation of a Sigma-sum involves, in an object-oriented implementation, different methods according to the different types of summands: these specially designed methods will take care of the properties of sigma which are needed, like convergence/divergence, index independence, constant summand and so on.

3 A specification of indexed symbolic objects, sums and series

The following specification of a system, for the manipulation of indexed mathematical objects in sums and series, translates the formalization given in the previous section. The specification language we use is tailored on the object-oriented view of software design and, besides the standard keywords sorts, operations, we allow, after the keyword inherit, the specification of inherited data types. In the new type, inherited operations are renamed so that the renaming hides the inherited definition. The keyword proper introduces operations that are newly defined. Furthermore, parametric specifications express polymorphic data types.

Domain describes the basic domains of computation, the set of basic types $\mathcal{D}_0$ and, used as parameter, it enriches the parameterized data type with properties derivable from the algebraic structure of the domain.

The simplest data type is the **Expression** type that consists only of a string. It is the supertype of the parametric type **Alg_Expression(Domain)**, and of the type **S_Expression** describing expressions containing Sigma signs. The entire hierarchy of data types below **Expression** is depicted in Fig. 1.

Alg_Expression(Domain) is the type of those expressions for which the algebraic properties of a specific algebraic domain hold. Typically, these are expressions that can be built by operations in the domain and can be rewrit-

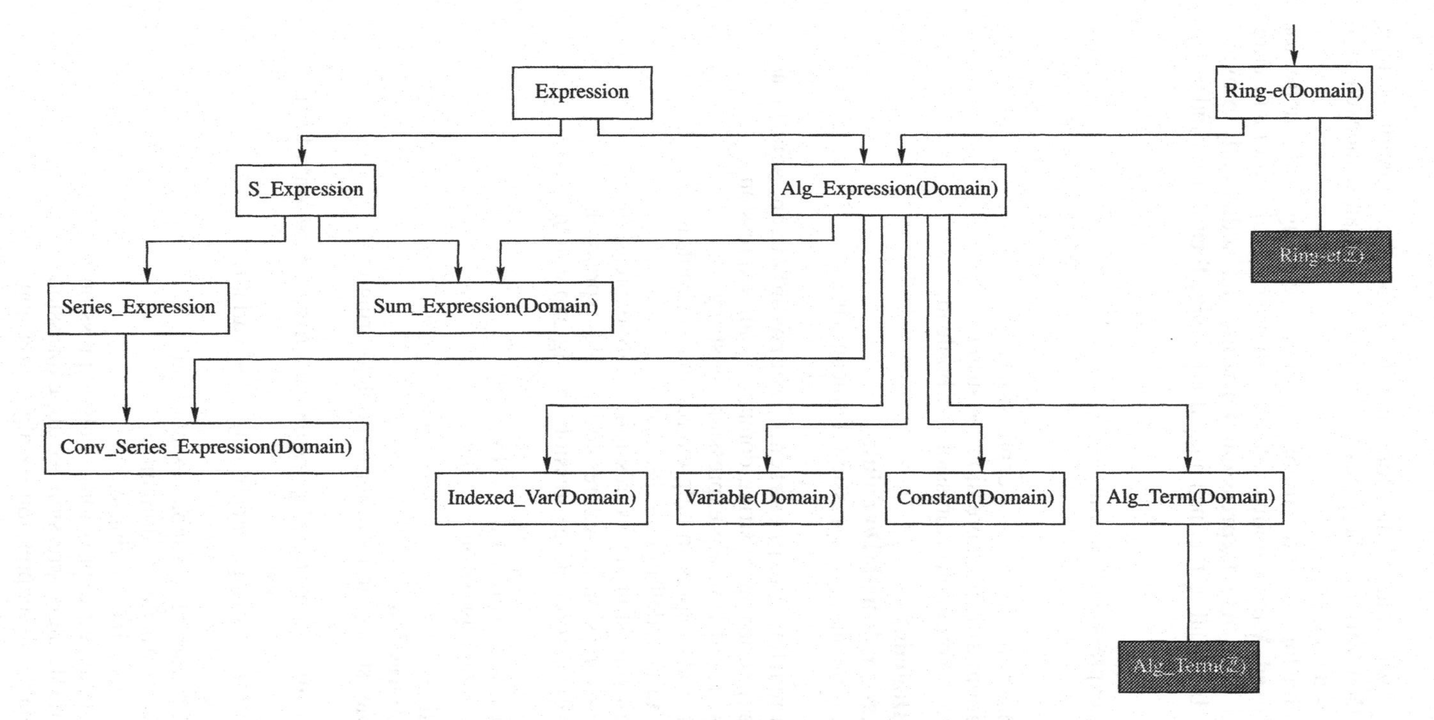

Fig. 1. Expression class/subclasses and class instances

ten according to laws valid in the domain. The data types, inheriting from **Alg_Expression(Domain)**, are the types for variables, constants, indexed variables, and terms of an algebra.

The specification for indexed variables is done according to the definition given above. **Indexed_Var(Domain)** redefines the name, value, domain components, inherited from **Alg_Expression(Domain)**, and additionally provides **Index**, defined to be equal to the instantiated type **Alg_Term($\mathbb{Z}$)**, as index.

Expression =
sorts string, expression, boolean;
operations
proper
  `expression_c`: string → expression
  `representation`: expression → string
  `equal?`: expression expression → boolean;

Alg-Expression(Domain) =
inherit **Expression, Ring-e(Domain)**;
sorts string, domain, alg-expression, boolean, ring-e;
operations
from **Expression** redefining `expression_c`, `representation`, `equal?` as:
  `alg-expression_c`: string domain → alg-expression
  `representation`: alg-expression → string
  `equal?`: alg-expression alg-expression → boolean
from **Ring-e(Domain)** redefining −, +, ∗ as:
  −: alg-expression alg-expression → alg-expression
  +: alg-expression alg-expression → alg-expression
  ∗: alg-expression alg-expression → alg-expression
proper
  `domain`: alg-expression → domain;
  `value`: alg-expression → ring-e;

Variable(Domain) =
inherit **Alg-Expression**;
sorts variable, string, domain, term, boolean, ring-e;
operations
from **Alg-Expression** redefining `alg-expression_c`, `representation`, `equal?`, `domain`, −, +, ∗, as:
  `variable_c`: string domain ring-e → variable
  `name`: variable → string
  `equal?`: variable variable → boolean
  `domain`: variable → domain
  `value`: variable → ring-e;
  −: variable alg-expression → alg-expression
  +: variable alg-expression → alg-expression
  ∗: variable alg-expression → alg-expression
proper
  `assign`: variable ring-e → variable;

Index = Alg-Expression($\mathbb{Z}$);

Indexed_Var(Domain) =
inherit **Alg-Expression**;
sorts indexed_var, index, string, domain, ring-e, boolean;
operations
from **Alg-Expression** redefining `alg-expression_c`, `representation`, `equal?`, `domain`, $-$, $+$, $*$, as:

`indexed_var_c`: string domain ring-e index → indexed_var
`representation`: indexed_var → string
`equal?`: indexed_var indexed_var → boolean
`domain`: indexed_var → domain
`value`: indexed_var → ring-e
$-$: indexed_var alg-expression → alg-expression
$+$: indexed_var alg-expression → alg-expression
$*$: indexed_var alg-expression → alg-expression

proper

`index`: indexed_var → index
`change_index`: indexed_var index index → indexed_var
`depend_on_index`: indexed_var index → boolean;

The **S_Expression** data type is characterized by a summand, that is an algebraic expression, and a range. It is a supertype for two subtypes: the type **Series-Expression** and the type **Sum-Expression** depending on whether the summation is infinite or finite. **S_Expression** elements are algebraic expression when the summations are either finite or convergent. In such cases, the manipulation rules that hold for ring operations can be inherited by these expressions too. We omit the specification of the ADT **Series-Expression** because it resembles that of **S-Expression**. The manipulation laws that we have recalled in the previous section are either defined for **S_Expression** or they are restricted to **Sum-Expression**.

S-Expression =
inherit **Expression**;
sorts expression, range, alg-expression, s-expression, boolean;
operations
from **Expression** redefining `expression_c`, `representation`, `equal?` as:

`s-expression_c`: index range alg-expression → s-expression
`representation`: s-expression → string
`equal?`: s-expression s-expression → boolean

proper

`summand`: s-expression → alg-expression
`index`: s-expression → index
`range`: s-expression → range
`distribute`: s-expression → alg-expression s-expression;
`split_domains`: s-expression s-expression → s-expression;

Sum-Expression(Domain) =
inherit **S-Expression, Alg-Expression(Domain)**;
sorts finite-range, alg-expression, sum-expression, boolean, ring-e;
operations
from **Alg-Expression** redefining `domain`, `value`, −, +, ∗ as:
    `domain`: sum-expression → domain
    `value`: sum-expression → ring-e
    −: sum-expression alg-expression → sum-expression
    +: sum-expression alg-expression → sum-expression
    ∗: sum-expression alg-expression → sum-expression
from **S-Expression** redefining `sigma_c`, `representation`, `equal?`, `summand`, `index`, `range`, `distribute`, `split_domains` as:
    `sum-expression_c`: index finite-range → sum-expression
    `representation`: sum-expression → string
    `equal?`: sum-expression sum-expression → boolean
    `summand`: sum-expression → alg-expression
    `index`: sum-expression → index
    `range`: sum-expression → finite_range
    `distribute`: sum-expression → alg-expression sum-expression;
    `split_domain`: sum-expression sum-expression → sum-expression;
proper
    `eval`: sum-expression → ring-e
    `interchange`: sum-expression sum-expression → sum-expression
    `change_index`: sum-expression index index → sum-expression;

Because these abstract specifications of data types are related to each other in a supertype/subtype relation, they can be directly implemented using an object-oriented environment as we will show in the next section.

4 Experimenting with CLOS

CLOS (Bobrow et al. 1988) is the object-oriented extension of Common Lisp based on the concepts of generic functions, multiple inheritance, and method combination. The underlying LISP substrate provides garbage collection, portability, and extendibility by symbolic manipulation facilities already available. Therefore, CLOS is a good tool for rapid prototyping. Since object-oriented design focuses on the abstract properties of objects rather than on implementation details, CLOS supports automatic control of the interactions among the objects by the dispatching mechanism associated to the generic functions. We briefly describe these attractive features of CLOS.

Generic functions are a generalization of ordinary Lisp function that replace the object-oriented mechanism of message passing. Given an operation and a tuple of objects on which to apply it, the most suitable code performing the operation is selected according to the classes of the objects. By the generic function approach, the operation is a generic function and the pieces of code are the *methods*. The methods define how to perform the operation on instances of specific classes. The generic function collects those methods and "dispatches" the right one. *Method combination* is the strategy that controls how to apply the applicable methods to a set of arguments in order to produce a value for the generic function.

Multiple inheritance is employed to build a type system of *classes* in which a class can inherit structure and behavior from a number of unrelated classes. It is a way of representing the relationships among various classes that enhances modularity and it avoids duplication of similar code.

A **class** is a Lisp object that defines structure and behavior for a set of other objects, called instances of that class. Classes are organized in a directed acyclic graph according to the inheritance relation. CLOS sets up a corresponding type for every class whose name coincides with that of the class.

A new class is defined using the macro **defclass** and it consists of a name, a list of direct super-classes, a set of slot specifiers and a set of class options. Slots are inherited by subclasses unless a new definition shadows them. The macro **make-instance** is used to create instances of a class. We have defined classes for the abstract data type specified in Sect. 3. Every class automatically contains a slot for each argument requested by its constructor function and, additionally, it contains functions to retrieve and modify each slot. As an example of CLOS code, the following shows that of the class meeting the specification of **Indexed_Ind(Domain)**.

```
(defclass indexed-var (alg-expression)
((name   :initarg :name
         :accessor name
         :type indexed-ind)
 (domain :initarg :domain
         :accessor id-domain)
 (value  :initarg :value
         :accessor id-value)
 (index  :initarg :index
         :accessor index
         :type index-i)))
```

A method is defined by invoking the **defmethod** macro. Methods implement class-specific behavior of generic functions. They are specified giving the list of parameter specializers, namely a list of classes. A method is selected when the argument satisfies the specializer list. In case of class-specializer, this means that the argument is an instance of that class or of one of its subclasses.

For example, the operation that renames an index has different behavior when the renaming is done on a **Sum_Expression** object or on an **Indexed_Var** object: in the first case, one has to check whether the index to be renamed is the dummy index or not. Therefore, two distinct methods implementing the function `change-index` are supplied.

```
(defmethod change-index ((oldindex index-i)
                         (newindex index-i)
                         (fs sum-expr))
```

```
  (setf (sum-range fs)
        (change-index oldindex newindex (sum-range fs)))
  (cond ((depend-on-index  oldindex (summand fs))
         (setf (summand fs)
               (change-index oldindex newindex (summand fs)))))
  fs)

(defmethod change-index ((oldindex index-i)
                         (newindex index-i)
                         (indexed indexed-var))
  (setf (index indexed)
        (change-index oldindex newindex (index indexed)))
  indexed)
```

Our implementation allows to manipulate, according to the rules that we have presented, finite sums in which the range is an interval.

5 Concluding remarks and future works

We have presented a study for the design of indexed objects, sums, and series as a proper example for testing the integrated system proposed by TASSO in Limongelli and Temperini (1992). As previously remarked, our aim is to define a set of rules to manipulate sums in the sense of manipulating either the summand or the relation over the index. At present, the general rules implemented are limited to finite sums and could be extended also to infinite sums, provided that convergence is ensured.

The current computer-algebra systems are usually equipped with a library of special case series like power series representations of functions and, to our knowledge, no available system includes a mechanism for testing convergence. It can be discussed whether a check of convergence should be included in the package in order to safely manipulate series. This would mean including some well known criteria for the convergence of series, such as D'Alembert or Cauchy's, whose application can be directed by the user.

We have carried out our experiments using CLOS but we are aware that this object-oriented programming environment still lacks some capabilities which would be useful in implementing designed mathematical data types. In particular, parametric classes could not be implemented in full generality.

Acknowledgement

This is the final report of a work done in 1991 in the frame of the project "Specification and Implementation of Methods for the Manipulation of Mathematical Objects for TASSO".

References

Bobrow, D. G., DeMichiel, L. G., Gabriel, P., Keene, S. E., Kiczales, G., Moon, D. A. (1988): Common Lisp object system specification. Tech. Rep. X3JI3 doc. 88-002R, ANSI Common Lisp Standard Committee. (SIGPLAN Not. 23/9).

Cioni, G., Miola, A. (1990): Specification and programming methodologies for axiomatizable objects manipulation: TASSO Project. Tech. Rep. P/04/31, Istituto di Analisi dei Sistemi ed Informatica, Consiglio Nazionale delle Ricerche, Rome.

Jenks, R. D., Trager, B. M. (1981): A language for computational algebra. In: Wang, P. (ed.): Proceedings ACM Symposium on Symbolic and Algebraic Manipulation, SYMSAC'81, Snowbird, Utah, USA, August 1981. Association for Computing Machinery, New York, pp. 6–13.

Limongelli, C., Temperini, M. (1992): Abstract specification of structures and methods in symbolic mathematical computation. Theor. Comput. Sci. 104: 89–107.

Miola, A. (1983): Design specifications for manipulation of sums and series. Tech. Rep. R.66, Istituto di Analisi dei Sistemi ed Informatica, Consiglio Nazionale delle Ricerche, Rome.

Shen, W., Wall, B., Wang, D. (1992): Manipulating uncertain mathematical objects: the case of indefinite sums and products. Tech. Rep. 92-23, RISC Linz, Johannes Kepler University, Linz.

Wang, D. (1990): Differentiation and integration of indefinite summations. Tech. Rep. 90-37.0, RISC Linz, Johannes Kepler University, Linz.

Wang, D. (1991): A toolkit for manipulating indefinite summations with application to neural networks. ACM SIGSAM Bull. 25/3: 18–27.

Programming methodologies

Equational specifications: design, implementation, and reasoning

S. Antoy, P. Forcheri, J. Gannon, and M. T. Molfino

1 Introduction

Sets of equations specify software systems either by describing the result of a computation or by characterizing some properties of the result. Suppose that the problem at hand is that of sorting a sequence of elements. We specify an operation *sort*, using an auxiliary operation *insert*, as follows:

$$
\begin{aligned}
&sort(nil) = nil \\
&sort(cons(E, L)) = insert(E, sort(L)) \\
&insert(E, nil) = cons(E, nil) \\
&insert(E, cons(F, L)) = \text{if } E \leq F \text{ then } cons(E, cons(F, L)) \\
&\quad \text{else } cons(F, insert(E, L))
\end{aligned}
$$

Here, we assume that sequences are constructed by "*nil*" and "*cons*" and that "$\leq$" denotes some ordering relation among the elements of a sequence. These equations specify the result of applying the operation *sort* to a sequence of elements.

An alternative approach consists in specifying properties of the result of applying the operation *sort* to a sequence of elements. There are two relevant properties. One is that of "being *sorted*," which is specified as follows:

$$
\begin{aligned}
&sorted(nil) = true \\
&sorted(cons(E, nil)) = true \\
&sorted(cons(E, cons(F, L))) = E \leq F \text{ and } sorted(cons(F, L))
\end{aligned}
$$

The other property of the result is that of "being a *permutation*" of the input. Informally, the sequences L and M are permutations of each other if and only if any element occurs the same number of times in L and M.

Both specifications are concerned with describing the input/output relation of the same problem, thus they can be easily related to each other:

$$sort(X) = Y \quad \text{if and only if} \quad sorted(Y) \text{ and } permutation(X, Y)$$

Although each specification is complete for the problem at hand and independent of the other, we regard them as complementary, rather than alternative. Concordance of independent specifications alleviates a problem occurring in the early phases of the software lifecycle. Decisions made during later phases, such as coding and testing, can be traced to decisions of earlier phases, e.g., a formal specification. However, early decisions, such as the specification itself, are based on information and knowledge which is seldom formalized. The checkable redundancy of the two forms of specification may allow us to catch some errors or to increase our confidence that a problem is well understood.

In addition to the above advantages, obvious differences in the two specifications may allow us to attack a problem from different angles. For example, the first specification is simple and direct, but may implicitly suggest an implementation of the sorting technique known as straight insertion. The second specification is more declarative, but does not immediately guarantee either the existence or uniqueness of sorting, and makes its implementation less obvious.

This contribution is concerned with the design, validation, use, and integration of different forms of specification. In Sect. 2 we discuss the framework and the notation we use for our specifications. In Sect. 3 we present some techniques for designing specifications with desirable properties of completeness and consistency. In Sect. 4 we describe the direct implementation of our specifications and uses of this implementation. In Sect. 5 we outline the structure of an automated prover for our class of specifications and its use for code verification.

Some aspects of our discussion have been treated by others, e.g., OBJ (Goguen and Winkler 1988) and Larch (Garland et al. 1990). Our approach emphasizes design and the dual role of specifications: description of behavior and description of properties. We address the integration of these roles and the application of the first to prototyping and testing, and the second to verification and reasoning.

2 Specification framework

We use an abstract language to present specifications. Lower case identifiers are symbols of the signature. If a signature is constructor-based, its constructors will be explicitly mentioned. The arity and co-arity of a symbol will generally be inferred from the context, or will be explicitly mentioned when non-obvious. Variables, which are sorted, are denoted by upper case letters. The variables in an equation are universally quantified. We are mainly concerned with the design and use of axioms; thus our specifications often consist only of equations.

Given a signature $\mathcal{F}$ and a set E of equations, the meaning of our syntactic constructs is given by the initial $\mathcal{F}$-algebra among all models of E. Reasoning based on the replacement of equals for equals is clearly sound, and it is complete in the sense of the Birkhoff theorem. We use rewriting (Dershowitz and Jouannaud 1990, Klop 1992) to restrict the freedom of replacing equals for equals, which is operationally very difficult to control. For convergent rewrite systems, equational reasoning is implemented by reduction to normal form without loss of reasoning power.

A traditional solution to employ rewrite rules instead of equations consists in transforming the equations of a specification into a corresponding rewrite system by means of the Knuth–Bendix completion procedure (Knuth and Bendix 1970). Although this approach appears simple to the specifier, the completion procedure may fail, may not terminate, or may require user intervention for ordering some rules it generates. We will discuss an alternative, based on design strategies, that yields a specification that is already in the form of a convergent rewrite system.

Our strategies require constructor-based signatures. This condition is not overly restrictive for the specification of software, but introduces the new problem of the sufficient completeness of the specification. We will address this problem, too. The signature of a specification is partitioned into constructors and (defined) operations when there exists a clear separation between data and computations. Every term built only from constructors represents a datum and every datum is represented by a constructor term. Every operation represents a computation and every computation is represented by a term containing an operation symbol. This viewpoint leads to considerable benefits discussed shortly. Referring to the examples of Sect. 1, *nil* and *cons* are the constructors of the type *sequence*, whereas *sort* and *sorted* are operations defined on the type *sequence*.

Equational reasoning may be too weak for proving all the interesting properties of a specification. Some properties may be provable via structural induction (Burstall 1969) or data type induction (Guttag et al. 1978). We employ the following principle. An inductive variable of type T is replaced by terms determined by T's constructors and inductive hypotheses are established. If F is a formula to be proved, v is the inductive variable, and s is the type of v, our induction proofs are carried out in the following manner. For every constructor c of type $s_1 \times \ldots \times s_n \to s$ for $n \geq 0$, we prove $F[c(v_1, \ldots, v_n)/v]$, where v_i, $1 \leq i \leq n$, is a distinct Skolem constant; and if $s_i = s$, then $F[v_i/v]$ is an inductive hypothesis.

3 Specification design

3.1 Completeness and consistency of definition

A goal of any specification is to avoid the errors of saying too much, which may lead to inconsistency, or too little, which may lead to incompleteness. Signatures with constructors help us to avoid these errors. A constructor-based specification should describe the behavior of an operation on every combination of constructor terms of the appropriate type and only on these terms. Ideally, for every combination of arguments of an operation there is one and only one axiom that defines the behavior of the operation on those arguments. A technique called the *binary choice strategy* (BCS) allows us to produce left sides of axioms with this property. The BCS is conveniently explained by means of an interactive, iterative, non-deterministic procedure that through a sequence of binary decisions generates the left sides of the axioms of a defined operation. We used the symbol "$\square$", called *place*, as a placeholder for a decision. Let f be an operation of type $s_1, \ldots, s_k \to s$. Consider the template $f(\square, \ldots, \square)$, where the i-th place has type s_i. To construct the set of left sides of rules for f, we replace, one at a time, each place of a template with either a variable or with a series of constructor terms of the appropriate type. In forming the left sides, we neither want to forget some combination of arguments, nor include other combinations twice. That is, we want to avoid both underspecification and overspecification. This is equivalent to forming a *constructor enumeration* (Choppy et al. 1989).

We achieve our goal by selecting a place in a template and choosing one of two options: *variable* or *inductive*. The choice *variable* replaces the selected place with a fresh variable. The choice *inductive* for a place of type s_i splits the corresponding template in several new templates, one for each constructor c of type s_i. Each new template replaces the selected place with $c(\square, \ldots, \square)$, where there are as many places as the arity of c. A formal description of the strategy appeared in Antoy (1990). The BCS produces the left sides of the rules of the operation *sorted* shown in Sect. 1 as follows. The initial template is:

$$sorted(\square)$$

We choose *inductive* for the place. Since this place is of type *sequence*, we split the template into two new templates, one associated with *nil* and the other with *cons*:

$$sorted(nil)$$
$$sorted(cons(\square, \square))$$

We now choose *variable* for the first place of the second template and then *inductive* for the second place to obtain:

$$sorted(nil)$$
$$sorted(cons(E, nil))$$
$$sorted(cons(E, cons(\square, \square)))$$

We choose *variable* for last remaining places and obtain the left sides of the axioms shown in Sect. 1.

If the constructors of the signature are free, the BCS produces axioms with non overlapping left sides and consequently avoids overspecification and inconsistencies in a specification. The BCS also produces completely defined operations. This, however, is not enough to avoid underspecification or to ensure that the specification is sufficiently complete. To achieve this other goal, we need the strategy described next.

3.2 Sufficient completeness and termination

If a specification is terminating and its operations are completely defined, as provided by the BCS, then the specification is also sufficiently complete. We use a technique called *recursive reduction strategy* (RRS) to guarantee the termination, and thus the sufficient completeness, of a specification. The RRS is conveniently explained by means of a function mapping terms to terms. The recursive reduction of a term t is the term obtained by "stripping" t of its recursive constructors. A constructor of type s is called *recursive* if it has some argument of type s. For example, *cons* is a recursive constructor of *sequence* because its second argument is of type *sequence*. Informally, "stripping" a term t is the operation of replacing any subterm of t rooted by a recursive constructor with its recursive argument. The stripping process is recursively applied throughout the term. A formal description of the recursive reduction function appears in Antoy (1990). We show its application in examples.

For reasons that will become clear shortly, we are interested in computing the recursive reduction of the left side of a rewrite rule for use in the corresponding right side. The symbol \$ in the right side of a rule denotes the recursive reduction of the rule's left side. With this convention, the second axiom of the operation *insert* is written:

$$insert(E, cons(F, L)) = \text{if } E \leq F \text{ then } cons(E, cons(F, L)) \text{ else } cons(F, \$)$$

since the recursive reduction of the left side is $insert(E, L)$. We obtain it by replacing $cons(F, L)$ with L, since L is the recursive argument of *cons*. Likewise, the second axiom of *sort* is written:

$$sort(cons(E, L)) = insert(E, \$)$$

When a constructor has several recursive arguments the recursive reduction of a term may require an explicit indication of the selected argument. We may also specify a partial, rather than complete, "stripping" of the recursive constructors. This would be appropriate for the operation *sorted*. In the third axiom, the recursive reduction of $sorted(cons(E, cons(F, L)))$ is $sorted(L)$, but we strip only the outermost constructor and in the right side we use $sorted(cons(F, L))$.

It is not difficult to see that the recursive reduction of a term t containing recursive constructors yields a term smaller than t. This property is the key to ensuring termination. We design a specification incrementally as follows. Suppose that S_i is a "well-behaved" specification, i.e., completely defined, consistent, terminating, and sufficiently complete. We extend S_i with some new operations using the previously discussed strategies. We obtain a new specification, S_{i+1}, that is still well behaved and is a conservative extension (in the sense of Ehrig and Mahr 1985) of S_i.

We show how our technique produces the first specification in Sect. 1. We assume that the operation "$\leq$", generic in our specification, is defined within a well-behaved specification. The "if-then-else" operation, which we consider a primitive, is well behaved too. The initial specification, S_0, consists of the free constructors of *sequence*. Since there are no axioms in S_0, completeness of definition, consistency, termination, and sufficient completeness are all trivially satisfied. Now we extend S_0 by defining the operation *insert*. We define its left sides with the BCS. The right side of each axiom is obtained by functional composition of symbols in S_0 and possibly *insert*, the operation we are defining. The latter can occur only within the recursive reduction of the left side, when some argument is a recursive constructor. This yields an extension of S_0 that we denote with S_1. The good behavior of S_0 and the use of the strategies ensure the good behavior of S_1. Now we can similarly extend S_1 with the operation *sort* and obtain a new well-behaved specification.

Our design approach is not appropriate for every specification. When it can be used, however, the approach yields a specification with desirable properties that are undecidable in general and not easy to verify in practice. If the signature contains non-free constructors, then to ensure the good behavior of the specification we must still verify its termination and consistency. For the latter it suffices to verify that every critical pair is convergent.

4 Direct implementation

4.1 Translation scheme

Rewriting is a model of computation. Specifications in the form of a rewrite system are executable – one simply rewrites terms to their normal forms. This implementation of a specification, known as *direct* (Guttag et al. 1978), is relatively straightforward for convergent, constructor-based systems. The interest in directly implementing a specification stems from the possibility of executing the specified software without incurring the cost of developing the code. That is, the direct implementation of the specification is a software prototype.

Many specification environments supporting rewriting, e.g., Garland et al. (1990) and Goguen and Winkler (1988), offer this form of prototyping. A common limitation of these prototypes is that they can be executed only in the specification environment. If the prototype is activated by existing code, or uses object libraries, or interacts with the operating system, then it may become necessary to execute the prototype in the same environment that will host the final code. A solution to this problem consists in mapping the rewrite system to various computational paradigms or particular languages (Antoy et al. 1990). For example, in functional and procedural languages constructors are mapped into code that builds instances of dynamic data structures, whereas operations are mapped into subprograms. The description of the transformation of a specification in Prolog (Clocksin and Mellish 1984) is discussed next. We choose this language because it is well-suited for coding harnesses of software prototypes and for creating complex, structured data that exercise these prototypes. Prolog is also well-suited for symbolic manipulation. Many ideas discussed in this note, including the automated prover described in the next section, have been implemented in this language (Antoy et al. 1993). Adding the direct implementation to these tools makes a rich environment for reasoning and experimenting with specifications.

If f is an operation with n arguments, the direct implementation in Prolog of f is a predicate `f` with $n+1$ arguments. The additional argument of `f` is used for "returning" the result of f applied to the other arguments. Each axiom defining an operation yields a Horn clause. In order to describe the details of the translation, we introduce a few notational conventions. We overload the comma symbol to denote both separation of string elements and concatenation of strings, i.e., if $x = x_1, \ldots, x_i$ and $y = y_1, \ldots, y_j$ are strings, with $i, j \geq 0$, then $x, y = x_1, \ldots, x_i, y_1, \ldots, y_j$. If τ is a function whose range is a set of non-null strings, then $\dot{\tau}(x)$ is the last element of $\tau(x)$, and $\bar{\tau}(x)$ is $\tau(x)$ without its last element. Combining the previous two notations, we have $\tau(x) = \bar{\tau}(x), \dot{\tau}(x)$.

The scheme for the direct implementation of a specification into Prolog is based on a function, τ, that maps terms of the specification into strings of Prolog terms. Symbols of the specification signature are mapped into Prolog symbols with the same spelling. The context of a symbol and the font in which is written, italic for the specification and typewriter for Prolog, resolve the potential ambiguity. T is a fresh Prolog variable.

$$\tau(t) = \begin{cases} \mathtt{X} & \text{if } t = X \text{ and } X \text{ is a variable,} \\ \bar{\tau}(t_1), \ldots, \bar{\tau}(t_k),\ \mathtt{c}(\dot{\tau}(t_1), \ldots, \dot{\tau}(t_k)) & \\ & \text{if } t = c(t_1, \ldots, t_k) \text{ and } c \text{ is a constructor,} \\ \bar{\tau}(t_1), \ldots, \bar{\tau}(t_k),\ \mathtt{f}(\dot{\tau}(t_1), \ldots, \dot{\tau}(t_k), \mathtt{T}),\ \mathtt{T} & \\ & \text{if } t = f(t_1, \ldots, t_k) \text{ and } f \text{ is an operation.} \end{cases}$$

Thus τ associates a Prolog predicate `f` to each operation f of the signature and an unevaluable symbol to each constructor. When it is extended from terms to axioms, τ yields Horn clauses.

$$\tau(f(t_1, \ldots, t_k) \to t) = \begin{cases} \mathtt{f}(t_1, \ldots, t_k, \dot{\tau}(t)) & \text{if } \bar{\tau}(t) \text{ is null,} \\ \mathtt{f}(t_1, \ldots, t_k, \dot{\tau}(t)) :- \bar{\tau}(t) & \text{otherwise.} \end{cases}$$

For example, to translate $\tau(sort(cons(E, L)) \to insert(E, sort(L)))$ to a Horn clause, we compute the following terms:

```
τ(insert(E,sort(L)))
    = τ̄(E),τ̄(sort(L)),insert(τ̇(E),τ̇(sort(L)),T),T
    = ϵ,sort(L,U),insert(τ̇(E),τ̇(sort(L)),T),T
    = sort(L,U),insert(τ̇(E),τ̇(sort(L)),T),T
    = sort(L,U),insert(E,U,T),T
τ(sort(L))
    = τ̄(L),sort(τ̇(L),U),U
    = ϵ, sort(L,U),U
    = sort(L,U),U
```

Thus $\tau(sort(cons(E, L)) \to insert(E, sort(L)))$ yields:

```
sort(cons(E,L),T) :- sort(L,U), insert(E,U,T)
```

An actual translator handles Prolog predefined types, such as numbers, and the "if-then-else" operation in an ad-hoc manner. The cut improves the efficiency of the directly implemented code. It avoids checking $E > F$ in the third clause of `insert` below. The direct implementation of our first specification is:

```
sort(nil,nil).
sort(cons(E,L),T) :- sort(L,U), insert(E,U,T).
insert(E,nil,cons(E,nil)).
insert(E,cons(F,L),cons(E,cons(F,L))) :- E<=F, !.
insert(E,cons(F,L),cons(F,T)) :- insert(E,L,T), .
```

This implementation scheme is equivalent to van Emden and Yukawa (1987), but the mapping τ provides more than a terse description of the transformation of a specification. An implementation of τ is retained in the prototype to provide a harness to invoke the directly implemented defined operations in a natural way. We define a predicate, `isab`, which plays, for abstract data types, the role played by the predefined predicate `is` for numeric types. `isab`, declared to be an infix operator with the same precedence and associativity of `is`, is (abstractly) implemented, using τ, as follows:

$$\texttt{isab}(\dot{\tau}(\texttt{X}),\texttt{X}) \; \texttt{:-} \; \texttt{call}(\bar{\tau}(\texttt{X})) \; .$$

For example, a harness for experimenting with the specifications discussed in Sect. 1 may attempt to verify that the result of sorting a sequence is sorted. The harness creates some sequence s and executes:

$$\texttt{N isab sorted(sort(}s\texttt{))} \; .$$

Expressions of this kind are simpler and more natural than the corresponding flattened expressions of van Emden and Yukawa (1987) and decrease the possibility of introducing errors in the harness.

4.2 Using prototypes

We use the prototypes obtained by the direct implementation in two ways. The previous section shows the first possibility. The prototype allows us to verify the agreement between our intuitive expectations and the behavior or properties described by the specification. For example, we verify whether *sort* applied to a sequence indeed sorts it, or whether *sorted* returns "true" if and only if its argument is a sorted sequence. We can also combine the two specifications, as shown in a previous example, to check their mutual agreement. Our specification is too simple to be of practical interest, but in principle there are no scalability problems.

When a prototype interacts with code not generated by the direct implementation of a specification, it is generated in the language of this code. Well-behaved specifications, such as those produced using the strategies discussed earlier, are easy to translate into imperative and functional languages too (Antoy et al. 1990).

The second possibility offered by the direct implementation of a specification originates from the use of the specification to check the correct execution of the code it specifies. For example, a C++ implementation of a class *sequence* has a method *sort* with the obvious meaning. The sortedness of a sequence object after sending it a *sort* message is a necessary condition for the correctness of the method. An assertion placed at the end of the method checks this condition:

```
void sequence::sort () {
  // body of the function
  assert (sorted (this object))
};
```

This use of specifications, discussed in Luckham (1990) for the Ada language and in Meyer (1988) for Eiffel, is effective, but has some drawbacks. Assertions involve methods of the class that may have to be coded just for this purpose, e.g., *sorted* or *permutation*. The same representation of *sequence* is used both by the operation being checked, *sort*, and by the operations checking it, *sorted*. Thus, loss of information in the representation may go undetected – for example the trivial type (which has a single value) satisfies any specification. Most important, it is not always easy to find properties that uniquely and completely characterize the behavior of an operation. What is missing in these approaches is a satisfactory degree of independence between the code and its assertions.

A more sophisticated approach (Antoy and Hamlet 1992) based on the direct implementation of a specification achieves this goal of independence. The specification is directly implemented with the intent of running it together with the code that it specifies. The direct implementation is no longer a prototype in the classical sense, but it coexists with the real code in the same run-time environment. The code uses the specification to check it itself. A significant advantage is using the specification of the behavior of a method, rather than the properties of its output. With some imprecision that we will correct shortly, the assertion for the *sequence* method *sort* becomes

```
assert (this object == spec_sort (this initial object))
```

where *spec_sort* is the name given to the direct implementation of the specification operation *sort* and *this initial object* is the state of the object before the execution of the *sort* method. Assertions of this kind are more convenient than those in Luckham (1990) and Meyer (1988) and do not require coding operations that appear only in the assertions. However, we must retain the state of the object at the time the method *sort* is invoked, and we must deal with two different representations for the type *sequence*: the representation chosen by the class implementer, that we refer to as concrete, and the representation generated by the direct implementation scheme, that we refer to as abstract.

The solution discussed in Antoy and Hamlet (1992) maintains both representations of an object. The test for equality performed by an assertion is between abstract representations. The concrete representation of an object at the end of the execution of a method is mapped to its abstract counterpart using what in Hoare (1972) is called a representation mapping. If we denote this mapping with $\mathcal{A}$, the required assertion for the example we are discussing is:

```
assert (A(this concrete object)
        == spec_sort (this initial abstract object))
```

The representation mapping $\mathcal{A}$ must be coded by the programmer. For example, suppose that the programmer represents a sequence using a dynamically allocated array. The instance variables of the class are the address of the array, the size of the array, and the number of elements in the sequence.

```
class sequence {
  private:
    int cursize;       // length of this sequence
    int maxsize;       // size of allocated array
    int * seq;         // address of array of elements
  //  Other private and public members
};
```

The representation mapping is shown below. The string "spec_" is prefixed to the symbols and the types directly implemented from the specification as opposed to those implemented by the programmer.

```
spec_sequence map (const sequence & s) {
  spec_sequence r = spec_nil ();
  for (int i = 0; i < s.cursize; i++) r = spec_cons (x[i], r);
  return r;
};
```

It is argued in Antoy and Hamlet (1992) that the extra programming effort required to code a representation mapping is a blessing in disguise. Unless the programmer has an accurate idea of this mapping, he cannot write correct methods that implement the specification's operations. There is no better way to ensure this understanding than to insist that it be put into code and executed.

The equality in the assertion is between abstract objects and thus corresponds to the syntactic equality of normal forms. The concrete representation of an object is maintained by the program. The abstract representation is computed within the assertion, thus only a minimum of bookkeeping is necessary to maintain it for use at a later time.

The second technique discussed in this section contains aspects of both multi-version programming, when the results of the directly implemented specification and its implementation by a programmer are compared for concordance, and self-checking, when the assertions check the internal state of an object. Both techniques have been investigated (Avizienis and Kelly 1984, Knight and Leveson 1986, Leveson et al. 1990) in relation to safety-critical applications.

5 Theorem proving

The most challenging application of a specification is to prove the correctness of software. We have experimented with an automated prover incorporating many concepts from the Boyer–Moore theorem prover (Boyer and Moore 1979). However, except for built-in knowledge of term equality and data type induction, the knowledge in the theorem prover is supplied by specifications. The prover takes advantage of the characteristics of a specification designed with the strategies that we have discussed earlier.

The theorem prover computes a boolean recursive function, called *prove*, whose input is an equation and whose output is "true" if and only if the equation has been proved. Axioms and lemmas of the specification are accessed as global data. Proofs of theorems are generated as side effects of computations of *prove*. By default, the prover autonomously performs reductions, selects inductive variables, generates the induction cases, applies the inductive hypotheses, and generalizes formulas to be proved. Users may override the automatic choices, made by the prover, for inductive variables and generalizations. A technique discussed in Antoy and Gannon (1994) allows users to employ a limited form of case analyses in proofs.

The theorem prover executes four basic actions for proving an equation: *reduce*, *fertilize*, *generalize*, and *induct*. *Reduce* applies a rewrite rule to the formula being proved. *Fertilize* is responsible for "using" an inductive hypothesis, i.e., replacing a subterm in the current formula with an equivalent term from an inductive hypothesis. *Generalize* tries to replace some non-variable subterm common to both sides of the formula with a fresh variable. *Induct* selects an inductive variable and generates new equations.

function $prove(E)$ is
  begin
    if E has the form $x = x$, for some term x, then return true; end if;
    if E can be reduced, then return $prove(reduce(E))$; end if;
    if E can be fertilized, then return $prove(fertilize(E))$; end if;
    $E' := generalize(E)$;
    if E' contains an inductive variable, then
      $E_1, \ldots, E_n := induct(E')$;
      return $prove(E_1)$ andalso ... andalso $prove(E_n)$;
    end if;
    return false;
  end *prove*;

An attempt to prove a theorem may exhaust the available resources, since induction may generate an infinite sequence of formulas to be proved. However, the termination property of the rewrite system guarantees that an equation cannot be reduced forever and the elimination of previously used inductive hypotheses (Boyer and Moore 1979) guarantees that an equation cannot be fertilized forever.

The strategies we have discussed ease some basic operations of the prover. The essential operations of every non-trivial proof involving an infinite type are *reduce* and *fertilize*. The BCS facilitates the first and the RRS the second.

The prover generates the formulas that constitute the cases of a proof by induction by instantiating an inductive variable with a constructor enumeration. When the inductive variable of a formula is an argument of an operation corresponding to an "*inductive*" choice in the operation's design with the BCS, the formulas that constitute the cases of a proof can all be immediately reduced.

The prover generates inductive steps of the form $l = r \supset l' = r'$. Fertilization is the process of replacing l with r in l' or, symmetrically, r with l in r'. If the inductive variable is selected according to the previous paragraph, then either l' or r' is reducible. If the axiom reducing one of these terms is designed with the RRS, then the reduced term can be fertilized.

An example

A common programming problem is the extraction of some information from a collection of values, for example, finding the smallest element. Let C be the collection and T the type of its elements. An informal specification of the above problem contains three steps. First, put aside one element of C, say f; second, find the smallest element, say g, of C without f; and third, return the smallest among f and g. This specification is just an instance of a computational paradigm known as *accumulation* (Basu 1980). An abstract, general accumulator is equationally specified by the function A defined next. The sequence $e_1, e_2, \ldots$ is a presentation in some order of the collection to be accumulated.

$$A(e_i, \ldots, e_j) = \begin{cases} \mathit{init} & \text{if } i > j \\ \mathit{step}(A(e_i, \ldots e_{j-1}), e_j) & \text{otherwise} \end{cases}$$

We can instantiate our example by defining the type *collection* with two constructors, *empty*, with the obvious meaning, and *add*, which takes an element e and a collection c and constructs the collection obtained by adding e to c. Replacing *step* with *min*, a function computing the minimum of two arguments, and *init* with *maxnat*, a special token discussed shortly, results in the following equations

$$\mathit{accum}(\mathit{empty}) = \mathit{maxnat}$$

$$\mathit{accum}(\mathit{add}(E, C)) = \mathit{min}(E, \mathit{accum}(C))$$

The symbol *maxnat* stands for the exception raised by the computation of the minimum of an empty collection. Order sorted specifications (Goguen 1978) give a precise meaning to this construction. The operation *min* is defined on a supersort of the natural numbers that contains *maxnat*, too. When one argument of *min* is *maxnat*, the operation returns the other argument. Thus, *min* sees *maxnat* as the "maximum natural number" and handles the exception. *Min* also propagates an exception if the other argument is *maxnat*, too.

Variants of this scheme can be used for a variety of other tasks, e.g., to add or multiply together the elements of the collection, when they are numbers, or to count how many elements satisfy a certain property, etc. The instantiations of *init* and *step* for a specific problem are generally easy to find, but require some care.

Our specification starts the accumulation from the initial portion of the presentation, as would be appropriate for a collection represented by an array. A different representation of the collection, e.g., a linked list, would suggest to start the accumulation from the other end. The second case of the definition of A would then be: $\mathit{step}(e_i, A(e_{i+1}, \ldots e_j))$.

While for our example the two versions are interchangeable; for other problems, such as converting a string of digits in the integer represented by the string, one version may be incorrect or inconvenient.

In general, if we want to prove the correctness of a piece of software or increase our confidence in the appropriateness of using an accumulator for a certain computation, we may have to reason about these definitions. For example, we may wish to prove that the two accumulations are equivalent or to verify whether the order in which the elements of the collection are accumulated affects the result of the accumulation.

The prover helps us accomplish these tasks. For example, we may prove that the equivalence of the above accumulation is implied by the following more general property:

$$A(e_i, \ldots, e_j) = step(A(e_i, \ldots, e_k), A(e_{k+1}, \ldots, e_j)) \quad i \leq k \leq j$$

A condition sufficient to ensure the validity of this equation is that $(T; step, init)$ is a left monoid. Thus, we may attempt to prove the above equations for some problem-specific instantiations of *init* and *step*.

In the appendix we discuss the specification of this problem in a form suitable for our prover and show the formal proofs that we obtain. The analysis of this problem suggests that it may be possible to implement some accumulations in parallel or by a divide-and-conquer technique. A further discussion on this opportunity appeared in Antoy and Gannon (1994).

6 Concluding remarks

We discussed some design techniques for an equational specification and some uses of the specification for software development. The specifications are in the form of constructor-based rewrite systems. The design techniques consist of two strategies and an incremental extension approach that allow us to produce specifications with desirable properties that are generally undecidable and difficult to verify in practice. The properties provided by our techniques are completeness of definitions, consistency, termination, and sufficient completeness. They also imply confluence and uniqueness of normal forms. The strategies are simple in practice and can be automated to a considerable degree. These strategies cannot be used for any specification, but when they can be used, they help in producing high quality specifications with relative ease.

Specifications with the properties provided by the strategies can be directly implemented in a variety of computing paradigms representative of mainstream programming languages. The direct implementation of a specification is modeled by rewriting. The resulting executable code can be activated by a harness that also provides test data and/or allows the inspection of the result of a computation. In this way we test whether our intuitive understanding of a problem is accurately captured by a formal specification.

A more sophisticated application allows a program to use a directly implemented specification as an oracle for self-checking. Self-checking is useful during the testing and debugging phases of the software lifecycle, since it minimizes oversights of the testing and debugging teams, it checks the internal states of objects in addition to their input/output relations, and it accurately localizes the region of a program where an error has occurred.

Specifications with the properties provided by the strategies facilitate reasoning about the specification themselves and about the correctness of programs. The BCS makes it easy to find inductive variables and to reduce all the cases of an induction. This, in conjunction with the RRS, creates in a formula the conditions for applying an inductive hypothesis. We cannot quantify the impact of these facts on proving theorems, although they appear to be quite useful.

Automated theorem proving is a very hard task. Our approach aims more at managing the complexity of a problem than on relying on the power of a prover. The theorem prover is conceptually simple. The printouts produced by the prover are easy to understand, and the user can follow the steps of a proof and the reasons of the prover for executing them. This is essential for discovering lemmas that help completing a difficult proof or for discovering that a specification does not capture the intuitive understanding of a problem. Although we can prove only very simple theorems, the prover expands our ability to reason about specifications and programs, and complements the opportunities provided by the direct implementation.

Appendix

We formalize the accumulator example discussed in the text as follows. The type *collection* has two constructors, *empty*, with the obvious meaning, and *add*, which takes an element e and a collection c and constructs the collection obtained by adding e to c.

Collection can be *concat*enated and *accum*ulated, where the definition of each operation is given by the axioms below. We label the axioms to reference their applications in a proof printout.

$concat(empty, C) = C$	concat_1
$concat(add(E, C), D) = add(E, concat(C, D))$	concat_2
$accum(empty) = init$	accum_1
$accum(add(E, C)) = step(E, accum(C))$	accum_2

The elements of a collection and the symbols *init* and *step* are generic, i.e., they are instantiated only for a specific problem, no constructors are defined for the type *element* and there are no defining axioms for the operations. We want to prove that if the type *element* with this operation is a left monoid, then the order in which the elements of a collection are accumulated does not affect the result of the accumulation. Hence, we assume:

$step(init, X) = X$	left identity
$step(step(X, Y), Z) = step(X, step(Y, Z))$	associativity

and we attempt to prove:

$$accum(concat(X, Y)) = step(accum(X), accum(Y))$$

The prover prints the following proof, where `A0`, `A1`, ... are the names of variables generated by the prover. Lines headed by "`IH-`" show inductive hypotheses, if any. Lines headed by "`(L)`" or "`(R)`" indicate the application of a transformation to the left or respectively right side of the equation being proved. The transformation is explained in the string delimited by "<<".

```
The theorem is:
  accum(concat(A0,A1)) = step(accum(A0),accum(A1))
Begin induction on A0
Induction on A0 case empty
Inductive hypotheses are:
  (L) accum(concat(empty,A1))  << subst A0 with empty <<
  (R) step(accum(empty),accum(A1))  << subst A0 with empty <<
  (L) accum(A1)  << reduct by concat_1 <<
  (R) step(init,accum(A1))  << reduct by accum_1 <<
  (R) accum(A1)  << reduct by left identity <<
  *** equality obtained ***
Induction on A0 case add(A2,A3)
Inductive hypotheses are:
  IH- accum(concat(A3,A1))=step(accum(A3),accum(A1))
  (L) accum(concat(add(A2,A3),A1))  << subst A0 with add(A2,A3) <<
  (R) step(accum(add(A2,A3)),accum(A1))  << subst A0 with
      add(A2,A3) <<
  (L) accum(add(A2,concat(A3,A1)))  << reduct by concat_2 <<
  (L) step(A2,accum(concat(A3,A1)))  << reduct by accum_2 <<
  (R) step(step(A2,accum(A3)),accum(A1))  << reduct by accum_2 <<
  (R) step(A2,step(accum(A3),accum(A1)))  << reduct by associativ-
      ity <<
  (L) step(A2,step(accum(A3),accum(A1)))  << ind. hyp. on A0 for
      A3 <<
  *** equality obtained ***
End induction on A0
QED
```

If we have an instance of an accumulation and, e.g., we want to implement it in parallel, it suffices to verify the left monoid property. Suppose that the problem is to find the minimum element of a collection. The type of the collection's elements is a subtype (supersort) of the natural numbers. The natural numbers are constructed by 0 and *succ* as usual. The supersort contains one extra element, *maxnat*. The axioms defining the operation *min* on this type are:

$min(0, X) = 0$	min 1
$min(succ(X), 0) = 0$	min 2
$min(succ(X), succ(Y)) = succ(\$)$	min 3
$min(succ(X), maxnat) = succ(X)$	min 4
$min(maxnat, X) = X$	min 5

Thus, we have to prove that *maxnat* is a left identity of *min* and that *min* is associative, i.e.:

$$min(maxnat, X) = X$$
$$min(min(X, Y), Z) = min(X, min(Y, Z))$$

The theorem prover prints, without assistance, the following proofs. The first is trivial. The second proof shows a triple nested induction.

```
The theorem is:
  min(maxnat,A0) = A0
  (L) A0  << reduct by min 5 <<
  *** equality obtained ***
QED

The theorem is:
  min(min(A1,A2),A3) = min(A1,min(A2,A3))
Begin induction on A1
Induction on A1 case 0
Inductive hypotheses are:
  (L) min(min(0,A2),A3)  << subst A1 with 0 <<
  (R) min(0,min(A2,A3))  << subst A1 with 0 <<
  (L) min(0,A3)  << reduct by min 1 <<
  (L) 0  << reduct by min 1 <<
  (R) 0  << reduct by min 1 <<
  *** equality obtained ***
Induction on A1 case succ(A4)
Inductive hypotheses are:
  IH- min(min(A4,A2),A3)=min(A4,min(A2,A3))
  (L) min(min(succ(A4),A2),A3)  << subst A1 with succ(A4) <<
  (R) min(succ(A4),min(A2,A3))  << subst A1 with succ(A4) <<
Begin induction on A2
Induction on A2 case 0
Inductive hypotheses are:
  IH- min(min(A4,A2),A3)=min(A4,min(A2,A3))
  (L) min(min(succ(A4),0),A3)  << subst A2 with 0 <<
  (R) min(succ(A4),min(0,A3))  << subst A2 with 0 <<
  (L) min(0,A3)  << reduct by min 2 <<
  (L) 0  << reduct by min 1 <<
  (R) min(succ(A4),0)  << reduct by min 1 <<
  (R) 0  << reduct by min 2 <<
  *** equality obtained ***
Induction on A2 case succ(A5)
Inductive hypotheses are:
  IH- min(min(A4,A5),A3)=min(A4,min(A5,A3))
  IH- min(min(succ(A4),A5),A3)=min(succ(A4),min(A5,A3))
  (L) min(min(succ(A4),succ(A5)),A3)  << subst A2 with succ(A5) <<
  (R) min(succ(A4),min(succ(A5),A3))  << subst A2 with succ(A5) <<
  (L) min(succ(min(A4,A5)),A3)  << reduct by min 3 <<
Begin induction on A3
Induction on A3 case 0
Inductive hypotheses are:
  IH- min(min(A4,A5),A3)=min(A4,min(A5,A3))
  IH- min(min(succ(A4),A5),A3)=min(succ(A4),min(A5,A3))
  (L) min(succ(min(A4,A5)),0)  << subst A3 with 0 <<
```

```
  (R) min(succ(A4),min(succ(A5),0))  << subst A3 with 0 <<
  (L) 0  << reduct by min 2 <<
  (R) min(succ(A4),0)  << reduct by min 2 <<
  (R) 0  << reduct by min 2 <<
  *** equality obtained ***
Induction on A3 case succ(A6)
Inductive hypotheses are:
  IH- min(min(A4,A5),A6)=min(A4,min(A5,A6))
  IH- min(min(succ(A4),A5),A6)=min(succ(A4),min(A5,A6))
  IH- min(succ(min(A4,A5)),A6)=min(succ(A4),min(succ(A5),A6))
  (L) min(succ(min(A4,A5)),succ(A6))  << subst A3 with succ(A6) <<
  (R) min(succ(A4),min(succ(A5),succ(A6)))  << subst A3 with
      succ(A6) <<
  (L) succ(min(min(A4,A5),A6))  << reduct by min 3 <<
  (R) min(succ(A4),succ(min(A5,A6)))  << reduct by min 3 <<
  (R) succ(min(A4,min(A5,A6)))  << reduct by min 3 <<
  (L) succ(min(A4,min(A5,A6)))  << ind. hyp. on  A1 for A4 <<
  *** equality obtained ***
Induction on A3 case maxnat
Inductive hypotheses are:
  IH- min(min(A4,A5),A3)=min(A4,min(A5,A3))
  IH- min(min(succ(A4),A5),A3)=min(succ(A4),min(A5,A3))
  (L) min(succ(min(A4,A5)),maxnat)  << subst A3 with maxnat <<
  (R) min(succ(A4),min(succ(A5),maxnat))  << subst A3 with maxnat <<
  (L) succ(min(A4,A5))  << reduct by min 4 <<
  (R) min(succ(A4),succ(A5))  << reduct by min 4 <<
  (R) succ(min(A4,A5))  << reduct by min 3 <<
  *** equality obtained ***
End induction on A3
Induction on A2 case maxnat
Inductive hypotheses are:
  IH- min(min(A4,A2),A3)=min(A4,min(A2,A3))
  (L) min(min(succ(A4),maxnat),A3)  << subst A2 with maxnat <<
  (R) min(succ(A4),min(maxnat,A3))  << subst A2 with maxnat <<
  (L) min(succ(A4),A3)  << reduct by min 4 <<
  (R) min(succ(A4),A3)  << reduct by min 5 <<
  *** equality obtained ***
End induction on A2
Induction on A1 case maxnat
Inductive hypotheses are:
  (L) min(min(maxnat,A2),A3)  << subst A1 with maxnat <<
  (R) min(maxnat,min(A2,A3))  << subst A1 with maxnat <<
  (L) min(A2,A3)  << reduct by min 5 <<
  (R) min(A2,A3)  << reduct by min 5 <<
  *** equality obtained ***
End induction on A1
QED
```

Acknowledgments

We would like to thank A. Miola for coordinating the TASSO project and all the members of the project for stimulating discussions. This work has been partially supported by CNR under the project "Sistemi Informatici e Calcolo Parallelo", grant no. 90.00750.69, by the National Science Foundation grants CCR-9196023, and CCR-9406751, and by the Office of Naval Research grants N00014-87-K-0307 and N0014-90-J4091.

References

Antoy, S. (1990): Design strategies for rewrite rules. In: Kaplan, S., Okada, M. (eds.): Conditional and typed rewriting systems. Springer, Berlin Heidelberg New York Tokyo, pp. 333–341 (Lecture notes in computer science, vol. 516).

Antoy, S., Gannon, J. (1994): Using term rewriting system to verify software. IEEE Trans. Software Eng. 20: 259–274.

Antoy, S., Hamlet, D. (1992): Automatically checking an implementation against its formal specifications. In: Shelby, E. R. W. (ed.): Proceedings 2nd Irvine Software Symposium, Irvine, CA, pp. 29–48.

Antoy, S., Forcheri, P., Molfino, M. T. (1990): Specification-based code generation. In: Shriver, B. D. (ed.): Proceedings 23rd Hawaii International Conference on System Sciences, vol. II, software. IEEE Computer Society Press, Los Alamitos, pp. 165–173.

Antoy, S., Forcheri, P., Molfino, M., Schenone, C. (1993): A uniform approach to deduction and automatic implementation. In: Fitch, J. (ed.): Design and implementation of symbolic computation systems. Springer, Berlin Heidelberg New York Tokyo, pp. 132–144 (Lecture notes in computer science, vol. 721).

Avizienis, A., Kelly, J. (1984): Fault tolerance by design diversity: concepts and experiments. Computer 17: 67–80.

Basu, S. (1980): On development of iterative programs from functional specifications. IEEE Trans. Software Eng. 6: 170–182.

Boyer, R., Moore, J. (1979): A computational logic. Academic Press, Orlando.

Burstall, R. (1969): Proving properties of programs by structural induction. Comput. J. 12: 41–48.

Choppy, C., Kaplan, S., Soria, M. (1989): Complexity analysis of term-rewriting systems. Theor. Comput. Sci. 67: 261–282.

Clocksin, W., Mellish, C. (1984): Programming in Prolog, 2nd edn. Springer, Berlin Heidelberg New York Tokyo.

Dershowitz, N., Jouannaud, J. (1990): Rewrite systems. In: van Leeuwen, J. (ed.): Handbook of theoretical computer science B: formal methods and semantics. North-Holland, Amsterdam, pp. 243–320.

Ehrig, H., Mahr, B. (1985): Fundamentals of algebraic specifications 1: equations and initial semantics. Springer, Berlin Heidelberg New York Tokyo.

Garland, S. J., Guttag, J. V., Horning, J. J. (1990): Debugging Larch shared language specifications. IEEE Trans. Software Eng. 16: 1044–1057.

Goguen, J. A. (1978): Order sorted algebras. Tech. Rep. 14, Computer Science Department, University of California Los Angeles.

Goguen, J. A., Winkler, T. (1988): Introducing OBJ3. Tech. Rep. SRI-CSL-88-9, SRI International, Menlo Park, CA.

Guttag, J. V., Horowitz, E., Musser, D. (1978): Abstract data types and software validation. Commun. ACM 21: 1048–1064.

Hoare, C. A. R. (1972): Proof of correctness of data representations. Acta Inf. 1: 271–281.

Klop, J. (1992): Term rewriting systems. In: Abramsky, S., Gabbay, D., Maibaum, T. (eds.): Handbook of logic in computer science. Oxford University Press, Oxford, pp. 1–112.

Knight, J. C., Leveson, N. G. (1986): An experimental evaluation of the assumption of independence in multi-version programming. IEEE Trans. Software Eng. 12: 96–109.

Knuth, D. E., Bendix, P. B. (1970): Simple word problems in universal algebras. In: Leech, J. (ed.): Computational problems in abstract algebra. Pergamon, Oxford, pp. 263–297.

Leveson, N. G., Cha, S. S., Knight, J. C., Shimeall, T. J. (1990): The use of self checks and voting in software detection: an empirical study. IEEE Trans. Software Eng. 16: 432–443.

Luckham, D. C. (1990): Programming with specifications: an introduction to ANNA, a language for specifying Ada programs. Springer, Berlin Heidelberg New York Tokyo.

Meyer, B. (1988): Object-oriented software construction. Prentice-Hall, Englewood Cliffs.

van Emden, M. H., Yukawa, K. (1987): Logic programming with equations. Logic Program. 4: 265–288.

On the algebraic specification of classes and inheritance in object-oriented programming

F. Parisi-Presicce and A. Pierantonio

1 Introduction

Recently, a number of formalisms has been proposed for the specification, analysis, and better understanding of the object-oriented methodology which encompasses different stages of the software design process such as analysis, design, and programming. The success of such a paradigm is achieved in the practical side of design of software systems since it allows for a better maintenance and enhances factors like reusability, extensibility, and compatibility of software artefacts. On the other hand, such a popularity conferred magical qualities to the word *object-oriented.* In fact, it seems that a system (but also a commercial product) in order to be considered good, needs to be called object-oriented. This can be viewed as a side-effect of a more general phenomenon, the confusion which was originally ruling the area. In fact, the terminology is quite ambiguous and very often the same term refers to intrinsically distinguished concepts. Furthermore, a lack of foundations did not help in keeping techniques which are different, e.g., software reuse and functional specialization, separated.

This research focuses on the modeling of concepts from the object-oriented programming with enough generality to give formal support to those techniques which made object-oriented so successful, i.e., *encapsulation*, *genericity*, and *inheritance*. Such a model has a twofold nature since it can be used to analyze existing systems as well as for specifying complex software systems.

The formalism is based on algebraic specifications in Ehrig and Mahr (1985), intended in the widest possible sense, although presented in their simplest form for clarity of exposition. The choice of the algebraic framework is dictated by the simplicity of its semantics and the modularity of its tools.

The model of class is intended to capture both the corresponding features in existing languages and suggestions borrowed from other areas (Ehrig and Mahr 1990). The class consists then of five parts: two interfaces, instance interface and class interface, for the two different roles of the class; another one, the import, requiring a producer for that class; a parameter part to model genericity; and an implementation part that includes the other four, in addition to the hidden features of the class. With such a model as basis, we are then able to distinguish between the notion of inheritance based on specialization and the notion of inheritance based on the reuse of code. We show that the two notions are related and that, in fact, the former is sufficient to express the latter.

The languages analyzed are those in which the class is the combination of a type concept and a module concept. Other languages don't fall in such a category but nevertheless they can still be analyzed with respect to other characteristics. The reason can be found in the fact that object-oriented programming is meant for the design of robust software, in which case the class is viewed as a module, or can be meant for modelling application domains, in which case a module represents a domain analysis discretized through a hierarchy of classes related via the specialization of the concepts they represent. In either case, the type concept deals with a specification of behavior. At the moment, in the space of languages we can find systems such as OOZE (Alencar and Goguen 1991) and the OBJ family, which are able to handle behavior in a proper way. One way to consider type and classes as separate concepts is to restrict the specification of behavior to signatures, and deal only with signature types. In the POOL programming language, a type specification is a signature with a list of property identifiers. This specification method is very simple but it can play an important role in the task of programming considering a property identifier as an abbreviation for a formal specification.

Another key notion in object-oriented methodology is inheritance, which allows the definition of new classes starting from variables and methods of old classes. But this definition of inheritance is not precise enough and it has been pointed out that there are different views of such a mechanism (Snyder 1986). Although both are called inheritance, there is no confusion between the idea of code sharing and the notion of functional specialization.

The BETA programming language gives the possibility to extend in a subclass the old methods defined in a superclass. The extension consists in a portion of code specified in the subclass. In general, in the other languages taken into consideration inheritance allows the redefinition of an old method, resulting in a specialization process without being able to say anything about correctness. This is particularly true for C++ and Smalltalk but also for Eiffel, even if it allows the specification of methods by pre- and post-conditions. In this framework, multiple inheritance can be considered only for sharing and combining code; otherwise the problem of name clashes has no easy solution. BETA doesn't allow multiple inheritance, but POOL does, by decoupling inheritance and subtyping.

Other aspects taken into account concern encapsulation, genericity and strict typing. We agree with Lehrmann Madsen et al. (1990) and Schaffer et al. (1986) that a programming language must offer a proper trade-off between flexibility and static checking in order to be useful. The object-oriented programming languages we have analyzed are C++, Eiffel, Trellis/Owl, POOL, BETA, Smalltalk, but also characteristics of other languages, such as OOZE and CLOS, will be mentioned.

In the following sections, we give the formal definition of our model of class and introduce the notions of specialization and reusing inheritance. These two relations show how specialization is different from code reusing. Then we show how to simulate the former by the latter via a virtual class. Two new relations between pairs of classes are presented, relations based on the use of the import interface and the parameter part. One relation indicates whether one class produces an instance interface which satisfies the constraints set forth in the parameter part of another class. The effect of substituting the class for the parameter is shown to be a clean (i.e., whose semantical effect can be predicted) way to obtain a new class which inherits by specialization from the old one. The other relation matches the producer of some methods to a potential consumer of those methods by finding a correspondence between the class interface of a class and the import interface of another class. The effect of matching the needed import with the provided class interface is again seen as a clean way to obtain a new class which inherits by implementation from the producer and by specialization from the consumer.

2 Preliminary notation

In this section, we briefly review some basic notions on the algebraic specifications; details can be found in Ehrig and Mahr (1985). A *signature* Σ is a pair (S, OP) where S is a set of *sorts* and OP a set of *constant* and *function symbols*; constant symbols are referred to as operation symbols of arity 0. A *pointed signature* is a signature $\Sigma = (S, OP)$ with a distinguished element of the set S of sorts denoted by $pt(\Sigma)$. By a Σ-*algebra* $A = (S_A, OP_A)$ of a signature $\Sigma = (S, OP)$ we mean two families $S_A = (A_s)_{s \in S}$ and $OP_A = (N_A)_{N \in OP}$, where A_s is a set for each $s \in S$, and $N_A\colon A_{s_1} \times \ldots \times A_{s_n} \to A_s$ is a function for each operator symbol $N\colon s_1 \ldots s_n \to s$ and each $s_1 \ldots s_n \in S^+$, $s \in S$ (for constant symbols $N\colon \to s$, $N_A \in A_s$). The set of all Σ-algebras is denoted by $Alg(\Sigma)$.

If $\Sigma_1 = (S_1, OP_1)$ and $\Sigma_2 = (S_2, OP_2)$ are signatures, a *signature morphism* $h\colon \Sigma_1 \to \Sigma_2$ is a pair of functions $(h^S\colon S_1 \to S_2, h^{OP}\colon OP_1 \to OP_2)$ such that for each $N\colon s_1 \ldots s_n \to s$ in OP_1 and $n \geq 0$ we have $h^{OP}(N)\colon h^S(s_1) \ldots h^S(s_n) \to h^S(s)$ in OP_2. A signature morphism $h\colon \Sigma_1 \to \Sigma_2$ induces a *forgetful functor*

$$V_h\colon Alg(\Sigma_2) \to Alg(\Sigma_1)$$

defined, for each Σ_2-algebra A'', by $V_h(A'') = A' \in Alg(\Sigma_1)$ with $A'_s = A''_{h^S(s)}$ for each $s \in S_1$ and $N_{A'} = h^{OP}(N)_{A''}$ for each $N \in OP_1$. A *pointed signature morphism* is a signature morphism $h\colon \Sigma_1 \to \Sigma_2$ such that $h^S(pt(\Sigma_1)) = pt(\Sigma_2)$. It is easy to check that pointed signature morphisms are closed under composition.

An *algebraic specification* $SPEC = (\Sigma, E)$ is a pair consisting of a signature Σ and a set E of (positive conditional) equations. If $SPEC_1 = (\Sigma_1, E_1)$ and $SPEC_2 = (\Sigma_2, E_2)$ are two algebraic specifications, a *specification morphism* $f\colon SPEC_1 \to SPEC_2$ is a signature morphism $f\colon \Sigma_1 \to \Sigma_2$ such that the translation $f^{\#}(E_1)$ of the equations of $SPEC_1$ is contained in E_2. A *pointed algebraic specification* is an algebraic specification with a pointed signature. A *pointed specification morphism* between pointed specifications is a pointed signature morphism f such that $f^{\#}(E_1) \subseteq E_2$.

For notational convenience, when $SPEC = (\Sigma, E)$ is a pointed specification the distinguished sort $pt(\Sigma)$ will be also denoted by $pt(SPEC)$.

The algebraic specifications and the specification morphisms form the category CATSPEC of algebraic specifications (Ehrig and Mahr 1985). The CATSPEC category is closed with respect to pushouts and pullbacks.

3 The class model

In this section we propose a class specification in order to model with generality the class notion present in the current object-oriented programming languages. Such a mathematical notion could be fruitfully exploited to better understand the main mechanisms and their interactions.

The importance of inheritance is widely recognized, but it is not the only peculiar feature of the object-oriented paradigm: encapsulation is considered as important (America 1990). This protection mechanism allows to trace a boundary between the implementation and the outside. The operations which can be invoked over the instances of a class are just those listed in the external interface of the class itself and any attempt at executing a private operation results in an error. The minimalization of the interdependencies of separately written software components and the reduction of the amount of implementational details are among the major benefits due to this technique. The concept of abstract data type is strengthened by the presence of an external interface because of the separation of the functionalities versus the implementation of an abstraction.

Unfortunately, inheritance can reduce the benefits of encapsulation. In fact accessing, in a subclass, inherited variables leaves the designer of the superclass unable to rename, reinterpret, or remove these variables. On the other hand, there are two categories of clients of a class, those who need to manipulate objects (the clients of the instances) and those who want to reuse the class in order to specialize it or just reuse its code (the clients of the class) (Snyder 1986). Thus, some languages have two external interfaces, one for each kind of clients, and they are defined incrementally since the interface for the class clients has a greater view than that for the instance clients. We will call these interfaces *instance* and *class interface*.

Usually, inheritance allows to arrange classes in specialization hierarchies through a *top-down* process. Classes can also be defined by reusing the code of another one or instances from other classes in a *bottom-up* process. If we allow an explicit import interface where some features, which are necessary to realize the external behavior of a class, are required, we can implement a class assuming the existence of some functionalities, ignoring which class will provide them. Unfortunately, most programming languages have constructs for defining modules and reuse them only by listing their names. This approach goes in the opposite direction of abstraction since we need to declare explicitly which is the supplier of a particular abstraction.

The class we have described in Sect. 2 is composed of a certain number of parts: a parameter part, an import interface, an instance, and a class export interface. All these components declare signatures and their properties.

Definition 1 (Class specification and semantics). A class specification C_{spec} consists of five algebraic specifications *PAR* (parameter part), EXP_i (instance interface), EXP_c (class interface), *IMP* (import interface), and *BOD* (implementation part) and five specification morphisms as in the following commutative diagram.

$$\begin{array}{ccccc} PAR & \xrightarrow{e_i} & EXP_i & \xrightarrow{e_c} & EXP_c \\ {\scriptstyle i}\downarrow & & & & \downarrow{\scriptstyle v} \\ IMP & & \xrightarrow[s]{\qquad\qquad} & & BOD \end{array}$$

The specification EXP_i, EXP_c, and *BOD* are pointed specifications, and e_c and v are pointed specification morphisms.

The semantics $SEM(C_{\text{spec}})$ of a class specification is the set of all pairs (A_I, A_{E_c}) of algebras, where $A_I = V_s(A_B)$ and $A_{E_c} = V_v(A_B)$ for some $A_B \in Alg(BOD)$. The pointed sort $pt(EXP_i)$ is called class sort.

Interpretation. Each of the five parts consists not only of signatures, but also of equations, which describe some of the properties of the operations. We distinguish among constructors, which are zero arity functions, methods, which have the form *meth*: $ss_1 \dots s_n \to s$, and functions *fun*: $ss'_1 \dots s'_n \to s'$. The interfaces EXP_i and EXP_c describe the external access functions and their behavior: the former describes the messages which can be sent to the objects that are instances of the class, while the latter contains the methods which can be used by other classes. The part of *BOD* not in EXP_c is hidden from other classes. The specification *BOD* describes an implementation of the exported methods using the ones provided by the *IMP* specification. The import specification *IMP* contains information about *what* is needed by *BOD* to implement EXP_c, but not *which* class can provide it: the latter task is provided by the interconnection mechanisms. The specification *PAR* models genericity, unconstrained if the specification consists of sorts only, constrained when the parameter is required to have operations satisfying certain properties.

Remark. The semantics chosen for a class specification is only one of a number of possibilities and the theory presented here can be developed by choosing one of the alternatives. One possibility is a functorial semantics (Parisi-Presicce and Pierantonio 1991) where the meaning of a class specification is a (functorial) transformation which takes a model (an algebra in this case) of *IMP* and returns a model (again an algebra) of EXP_c. In the latter case, the semantics must be definable uniformly for all classes and must be closed with respect to amalgamation (Ehrig and Mahr 1985). Another possible choice is of a loose semantics, where only some pairs (A_I, A_{E_c}) are chosen.

Definition 2 (Class). A class $C = (C_{\text{spec}}, C_{\text{impl}})$ consists of a class specification

C_{spec} and a class implementation C_{impl} such that $C_{\text{impl}} = A_B$ for some *BOD*-algebra A_B.

Example 1. The morphisms in this example are just inclusions. In the notation we use the keywords **Parameter, Instance Interface, Class Interface, Import Interface, Body** to declare the subspecification to be added to the parts already defined. For example, since $PAR \subseteq EXP_i$, after the keyword **Instance Interface** only $EXP_i - PAR$ is listed. In the instance interface the distinguished sort $pt(EXP_i)$ is stated by the *class sort* clause.

Next, the FList class specification is defined. The elements of this abstract data type are *frequency lists*, i.e., lists which allow to count how many occurrences of a given element are stored in the list.

FList **is Class Specification**
Parameter
sort data
operations $\bot$: → data
Instance Interface
class sort flist
sort nat
operations 0: → nat
SUCC: nat → nat
NIL: → flist
(_): data → flist
TAIL: flist → flist
HEAD: flist → data
_ + _: flist flist → flist
FREQUENCY: flist data → nat
equations $\text{NIL} + l = l = l + \text{NIL}$
$\text{HEAD}((d) + l) = d$
$\text{TAIL}((d) + l) = l$
$\text{HEAD}(\text{NIL}) = \bot$
$\text{TAIL}(\text{NIL}) = \text{NIL}$
$\text{FREQUENCY}(\text{NIL}, d) = 0$
$\text{FREQUENCY}((d) + l, x) =$
if $d = x$ then $\text{SUCC}(\text{FREQUENCY}(l, x))$
else $\text{FREQUENCY}(l, x)$
Import Interface
sorts nat, string
operations 0: → nat
SUCC: nat → nat
EMPTY: → string
MAKE: data → string
CONCAT: string string → string
equations $\text{CONCAT}(s, \text{NIL}) = s$
$\text{CONCAT}(\text{CONCAT}(s1, s2), s3) =$
$\text{CONCAT}(s1, \text{CONCAT}(s2, s3))$

Body

operations $[\_]$: string $\rightarrow$ flist
$\text{NIL} = [\text{EMPTY}]$
$\text{HEAD}([\text{CONCAT}(\text{MAKE}(a), s)]) = a$
$\text{TAIL}([\text{CONCAT}(\text{MAKE}(a), s)]) = [s]$

End FList

The FList example also shows the role played by each component of the class specification. The instance interface describes the *abstract* properties, in the sense that the intended behavior is representation independent. The body describes, in turn, how such data type is implemented in terms of what is provided by the formal import. In other words, we suppose the imported items as predefined and then we use them for *programming* the features of the class. In this example, the class interface coincides with the instance one.

Presenting this class model we intend to cover a large number of class structures as they are defined in current object-oriented languages. None of the languages analyzed allows to specify some requirements for the import, although some allow the direct importing of other existing classes, incorporating (with the *use* clause) a combination mechanism. The opportunity to hide some implementational aspects gives to a class designer the freedom to modify the implementation without affecting the clients of the instances of that class. All the languages analyzed but BETA have constructs for protection of data representation. The set of all public operations of a class forms the external interface, which we call *instance interface*. Another form of protection is given to prevent another kind of client, the designer of a subclass, to access some variables. We have named this other interface, which contains the instance one, *class interface*. The C++, POOL, Trellis/Owl have an explicit class interface, distinct from the instance interface. For instance, in the C++ language the instance interface consists of all public items which can be declared like that via a *public* clause, while the class interface includes both the public and the subclass visible items, declared through a *protected* clause, accessible only to derived classes.

Encapsulation and inheritance are the major features of object-oriented methodology but other techniques can as well enhance some quality factors. Meyer (1986) presented an informal comparison between genericity and inheritance. Many languages, such as Eiffel, Trellis/Owl, POOL, BETA, and OOZE, allow genericity, although with some differences. The only properties treated by these languages are signature properties; OBJ allows one to specify behavior with an equational language, by means of theories and views, while OOZE uses pre- and postconditions in the style of Z. All these languages supply an actualization mechanism in order to instantiate the generic classes.

4 Inheritance

Inheritance is one of the main notions of the object-oriented paradigm. Its importance is widely recognized as it allows to reuse, extend, and combine abstractions in order to define other abstractions. It allows the definition of new classes starting from the variables and methods of other classes. The usual terminology calls the former *subclasses* and the latter ones *superclasses*. Unfortunately in the space of languages the notion of inheritance is not homogeneous since it ranges from functional specialization to the reuse of code without any constraint. We can consider inheritance as a technique for the implementation of an abstract data type: its use is a private decision of the designer of the inheriting class and the omission and/or shadowing of features is allowed. Through this mechanism we can arrange classes in hierarchies which describe how programs are structured: we call this technique *reusing inheritance*. On the other hand, we can consider inheritance as a technique for defining behavioral specialization: its use is a public declaration of the designer that the instances of the subclass obey the semantics of the superclass. Thus each subclass instance is a special case of superclass instance: we call this kind of inheritance *specialization inheritance*. The hierarchies obtainable by means of the specialization inheritance contain two kinds of information. On the one side they describe how code is distributed among classes and thus how programs are structured. On the other side they produce compatible assignment rules which are related with the subtyping relation: each context which expects a superclass instance can accept a subclass instance since the behavior of the subclass is at least that of the superclass.

We now present a formal distinction between these two different notions

Definition 3 (Reusing inheritance). Let $C1 = (C1_{\text{spec}}, C1_{\text{impl}})$ and $C2 = (C2_{\text{spec}}, C2_{\text{impl}})$ be classes. Then

1. $C2$ weakly reuses $C1$, notation $C2$ `Wreuse` $C1$, if there exists a morphism

$$f\colon EXP_{c1} \to BOD_2 ,$$

(called reusing morphism) as in Fig. 1, such that

$$f^S(pt(EXP_{c1})) = v_2^S(pt(EXP_{c2})) ;$$

2. $C2$ strongly reuses $C1$, notation $C2$ `Sreuse` $C1$, if, in addition,

$$V_f(C2_{\text{impl}}) = V_{v1}(C1_{\text{impl}}) .$$

Example 2. In this example, we introduce an ordinary list with an additional EMPTY? operation which checks whether the list is empty or not. The reusing inheritance allows us to define such an abstract data type starting from FList. Actually, we remove as visible attributes the FREQUENCY operations, which remain anyway in the body of FList, and introduce the new operation. The reusing specification morphism

$$f\colon FList.EXP_c \to EList.BOD$$

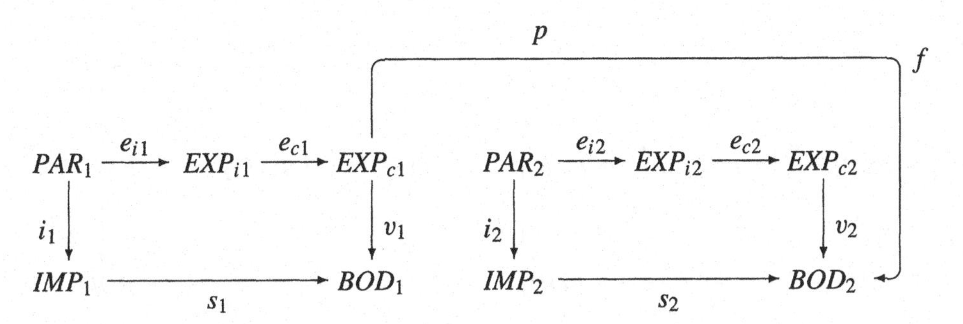

Fig. 1. Reusing inheritance

renames the sort flist into elist and changes the arity of the operation symbols accordingly.

Notice how the inherited features are defined from scratch in the body of the derived class while, for instance, the naturals are required in the formal import.

EList **is Class Specification**
Parameter

sort	data
operations	$\perp$: $\rightarrow$ data

Instance Interface

class sort	elist
sort	bool
operations	TRUE, FALSE: $\rightarrow$ bool
	NIL: $\rightarrow$ elist
	(_): data $\rightarrow$ elist
	TAIL: elist $\rightarrow$ elist
	HEAD: elist $\rightarrow$ data
	_ + _: elist elist $\rightarrow$ elist
	EMPTY?: elist $\rightarrow$ bool
equations	NIL $+ l = l = l +$ NIL
	HEAD$((d) + l) = d$
	TAIL$((d) + l) = l$
	HEAD(NIL) $= \perp$
	TAIL(NIL) = NIL
	EMPTY?(NIL) = TRUE
	EMPTY?$((d) + l)$ = FALSE

Import Interface

sorts	bool, nat
operations	TRUE, FALSE: $\rightarrow$ bool
	0: $\rightarrow$ nat
	SUCC: nat $\rightarrow$ nat

Body

operations	FREQUENCY: elist data $\rightarrow$ nat

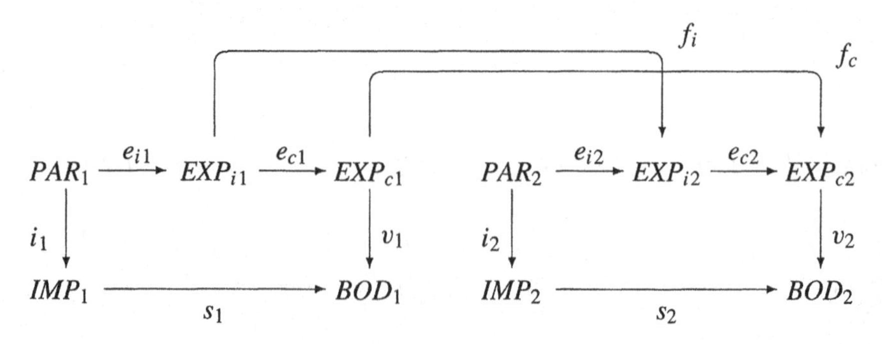

Fig. 2. Specialization inheritance

equations FREQUENCY(NIL, d) = 0
FREQUENCY((d) + l, x) =
if $d = x$ then SUCC(FREQUENCY(l, x))
else FREQUENCY(l, x)

End EList

The second notion of inheritance is that of specialization inheritance which allows the enrichment of the functionalities of a class and can be modeled by morphisms from the inherited class to the inheriting class in such a way that behavior is preserved.

Definition 4 (Specialization inheritance). Let $C1 = (C1_{\text{spec}}, C1_{\text{impl}})$ and $C2 = (C2_{\text{spec}}, C2_{\text{impl}})$ be classes. Then

1. $C2$ is a weak specialization of $C1$, notation $C2$ `Wspec` $C1$, if there exist pointed morphisms
$$f_i\colon EXP_{i1} \to EXP_{i2} \ ,$$
$$f_c\colon EXP_{c1} \to EXP_{c2} \ ,$$
(called specialization morphisms) such that $e_{c2} \circ f_i = f_c \circ e_{c1}$, as in the commutative diagram of Fig. 2;
2. $C2$ is a strong specialization of $C1$, notation $C2$ `Sspec` $C1$, if, in addition,
$$V_{f_c}(V_{v2}(C2_{\text{impl}})) = V_{v1}(C1_{\text{impl}}) \ .$$

Example 3. In order to illustrate the specialization inheritance we consider the following CFList class specification, which is a specialization of the FList in the sense of the previous definition. The specialization morphism as for the reusing inheritance renames the sort flist into cflist. The CFList introduces a count operation which returns the number of elements stored in the list.

CFList **is Class Specification**
Parameter
sort data

operations $\perp: \to$ data
Instance Interface
class sort cflist
sort nat
operations $0: \to$ nat
SUCC: nat $\to$ nat
NIL: $\to$ cflist
$(\_)$: data $\to$ cflist
TAIL: cflist $\to$ cflist
HEAD: cflist $\to$ data
$\_+\_$: cflist cflist $\to$ cflist
FREQUENCY: cflist data $\to$ nat
COUNT: cflist $\to$ nat
equations $\mathrm{NIL} + l = l = l + \mathrm{NIL}$
$\mathrm{HEAD}((d) + l) = d$
$\mathrm{TAIL}((d) + l) = l$
$\mathrm{HEAD}(\mathrm{NIL}) = \perp$
$\mathrm{TAIL}(\mathrm{NIL}) = \mathrm{NIL}$
$\mathrm{FREQUENCY}(\mathrm{NIL}, d) = 0$
$\mathrm{FREQUENCY}((d) + l, x) =$
if $d = x$ then $\mathrm{SUCC}(\mathrm{FREQUENCY}(l, x))$
else $\mathrm{FREQUENCY}(l, x)$
$\mathrm{COUNT}(\mathrm{NIL}) = 0$
$\mathrm{COUNT}((d) + l) = \mathrm{SUCC}(\mathrm{COUNT}(l))$
Import Interface
sorts nat
operations $0: \to$ nat
SUCC: nat $\to$ nat
End CFList

Remark. In general, the inheritance at the specification level is twofold since it can be instantiated at the specification level and at the code level. For instance, in the specialization inheritance the incremental requirement of properties and/or operations in the subclass implies the refinement of the set of all models of the superclass. This is captured by the weak specialization and it is mainly intended to support monotonic decision steps in the software design process. On the other hand, each of the models of the superclass may be related with a specific model of the subclass which is a subalgebra of the superclass model. Such a pointwise relation among algebras (abstract implementations) corresponds to the inheritance relation at the programming language level.

Although both are called inheritance, there is no confusion between the idea of code sharing and the notion of functional specialization. Notice how in general reusing inheritance does not give rise to correctness problems when its use is intended. Unfortunately, its use can be also non intended. Indeed, although the redefinition of methods may be useful to refine inherited operations, it can be dangerous since the redefinition can violate the invariants of the superclass resulting in something with an unexpected behavior. The intended specialization inheritance degenerates in reusing inheritance while the designer is expecting to use the subclass as a subtype, i.e., as something with a compatible behavior.

This is particularly true for C++ and Smalltalk but also for Eiffel, even if it allows the specification of the methods by pre- and postconditions. The BETA programming language does not allow the redefinition of methods and gives the possibility to extend in a subclass the old method defined in a superclass. The extension consists of a portion of code specified in the subclass. When a message is received by a class instance, the method executed is from the top-most superclass containing the message; the execution can, in turn, trigger another execution (by the imperative *inner*) from a subclass, and so on. Unfortunately there is no constraint in providing the methods extension and this could violate the invariants of the superclass as well.

The two inheritance relations satisfy some properties formalized in the following propositions.

Proposition 1. The strong relations `Sreuse` and `Sspec` imply the weak relations `Wreuse` and `Wspec`, respectively. Furthermore, each of the two relations `Sspec` and `Wspec` is transitive.

Proposition 2. Let $C1 = (C1_{\text{spec}}, C1_{\text{impl}})$, $C2 = (C2_{\text{spec}}, C2_{\text{impl}})$ and $C3 = (C3_{\text{spec}}, C3_{\text{impl}})$ be classes.

1. if $C2$ `Wspec` $C1$ then $C2$ `Wreuse` $C1$;
2. if $C2$ `Sspec` $C1$ then $C2$ `Sreuse` $C1$;
3. if $C3$ `Wreuse` $C2$ and $C2$ `Wspec` $C1$ then $C3$ `Wreuse` $C1$;
4. if $C3$ `Sreuse` $C2$ and $C2$ `Sspec` $C1$ then $C3$ `Sreuse` $C1$.

Another important notion is that of *virtual class*[1] which is formalized in the following definition.

Definition 5. A virtual class V consists of a triple (PAR, EXP_i, EXP_c) of algebraic specifications, a specification morphism e_i: $PAR \to EXP_i$ and a pointed specification morphism e_c: $EXP_i \to EXP_c$ as in the Fig. 3.

The following result shows that it suffices to consider specialization inheritance as the only relation generating a hierarchy of classes.

$$PAR \xrightarrow{e_i} EXP_i \xrightarrow{e_c} EXP_c$$

Fig. 3. Virtual class

1 Virtual class is not intended in the sense of the BETA language where it represents the parameter part of a class.

Theorem 1. Let $C1 = (C1_{spec}, C1_{impl})$ and $C2 = (C2_{spec}, C2_{impl})$ be classes. If $C2$ `Wreuse` $C1$ then there exists a virtual class C such that
1. $C1$ `Wspec` C;
2. $C2$ `Wspec` C.

By using the specialization inheritance (which implies subtyping) it is possible to represent the implementation inheritance (which usually reflects the development over time of the system) and keep the correctness under control. On the other hand, the theorem allows us to consider only monotonic decision steps (via specialization inheritance) since omitting properties and/or operations, i.e., non monotonic steps can be reduced to monotonic ones.

Example 4. We already know how the EList can be implemented from FList via reusing inheritance. The theorem assures the existence of a class, say VList, which is (weakly) specialized by EList and FList.

VList **is Class Specification**
Parameter
sort data
operations $\perp: \rightarrow$ data
Instance Interface
class sort vlist
sort bool
operations TRUE, FALSE: $\rightarrow$ bool
NIL: $\rightarrow$ elist
(_): data $\rightarrow$ elist
TAIL: vlist $\rightarrow$ elist
HEAD: vlist $\rightarrow$ data
_ + _: vlist vlist $\rightarrow$ vlist
equations $\text{NIL} + l = l = l + \text{NIL}$
$\text{HEAD}((d) + l) = d$
$\text{TAIL}((d) + l) = l$
$\text{HEAD}(\text{NIL}) = \perp$
$\text{TAIL}(\text{NIL}) = \text{NIL}$
End VList

The specialization morphism and the morphisms f'_P and f''_P are inclusions.

5 Structured inheritance

In this section, we define two additional relations between class specifications and between classes (weak and strong relations, respectively) which describe mechanisms for interconnecting class specifications and define new classes, which inherit from the old ones. The two relations correspond to instantiating the parameter part of a (generic) class and to replacing the import interface of a class with the class export of another class.

Definition 6 (Actualizable). Given classes $C1 = (C1_{\text{spec}}, C1_{\text{impl}})$ and $C2 = (C2_{\text{spec}}, C2_{\text{impl}})$ with $PAR_2 = IMP_2$.

1. $C1$ is weakly actualizable by $C2$, denoted by $C1$ `Wact` $C2$, if there exists a specification morphism $f\colon PAR_1 \to EXP_{i2}$;
2. $C1$ is strongly actualizable by $C2$, denoted by $C1$ `Sact` $C2$, if $C1$ `Wact` $C2$ with f and $V_f(V_{e_{c2}}(V_{v2}(C2_{\text{impl}}))) = V_{e_{i1}}(V_{e_{c1}}(V_{v1}(C1_{\text{impl}})))$.

Interpretation. The binary relation `Wact` indicates that the parameter part of the class $C1$ can be replaced by the instance interface of the class $C2$, i.e., that the instances of class $C2$ satisfy the constraints of the parameter of $C1$. The distinction between weak and strong again separates the class specification from the class. In $C1$ `Wact` $C2$, a realization of the class specification $C2_{\text{spec}}$ *can* be used for the parameter part of a realization of the class specification $C1_{\text{spec}}$. On the other hand, $C1$ `Sact` $C2$ indicates that the realizations $C1_{\text{impl}}$ and $C2_{\text{impl}}$ of the class specification *coincide* on their PAR_1 part.

The restriction imposed on the class $C2$ that $PAR_2 = IMP_2$ is both technical and methodological. On the technical side, it makes the construction of the resulting class specification after replacing the parameter part cleaner. On the methodological side, it requires that only *complete* classes, i.e., classes which do not rely on other classes for their completion of the import specification, be considered as actual parameters of generic classes.

We now describe the result of actualizing the parameter part PAR_1 of $C1$ by $C2$. The categorical constructions and the detailed proofs of syntactical and semantical properties are omitted.

Definition 7 (Parameter passing). Let $Ci = (Ci_{\text{spec}}, Ci_{\text{impl}})$, $i = 1, 2$, be classes with $C1$ `Wact` $C2$ via $f\colon PAR_1 \to EXP_{i2}$. The actualization of $C1$ by $C2$, denoted by $ACT(C1, f, C2)$, is the class specification $C3_{\text{spec}}$ where

- the parameter part PAR_3 is the parameter part PAR_2 of $C2$;
- the instance interface EXP_{i3} is the union of EXP_{i1} and EXP_{i2} (i.e., the pushout of f and e_{i1});
- the class interface EXP_{c3} is the union of EXP_{c1} and EXP_{c2} (i.e., the pushout of $f \circ e_{c2}$ and $e_{i1} \circ e_{c1}$);
- the import interface IMP_3 is the union of IMP_1 and PAR_2 (i.e., the pushout of the two with respect to the intersection of PAR_1 and PAR_2);
- the implementation part BOD_3 is the union of BOD_1 and BOD_2 with respect to PAR_1.

The specification morphisms are induced by the universal properties of the pushout objects.

The distinguished sorts of EXP_{i3}, EXP_{c3} and BOD_3 are the ones inherited from EXP_{i1}, EXP_{c1} and BOD_1, respectively, in the union construction.

Interpretation. The new class specification $ACT(C1, f, C2)$ is obtained by *re-*

placing the parameter part PAR_1 in EXP_{i1} and EXP_{i2} with the instance and class interface, respectively, of $C2_{spec}$, and in BOD_1 with the implementation part BOD_2. The new parameter part is just the parameter PAR_2 of $C2$, which is also added to IMP_1 to obtain the new import interface.

Theorem 2 (Induced semantics). The semantics of the class specification $ACT(C1, f, C2)$ is the set of all pairs $(A_{I_3}, A_{E_{c3}})$ such that:

$A_{I_3} = A_{I_2} +_{A_P} A_{I_1}$

$A_{E_{c3}} = A_{E_{c2}} +_{A_{P_1}} A_{E_{c1}}$

$(A_{I_1}, A_{E_{c1}}) \in SEM(C1_{spec})$

$(A_{I_2}, A_{E_{c2}}) \in SEM(C2_{spec})$

$V_{e_{i1}}(V_{e_{c1}}(A_{E_{c1}}) = V_f(V_{e_{c2}}(A_{E_{c2}}))$

$V_{p_2}(V_{i2}(A_{I_2})) = V_{p_1}(V_{i1}(A_{I_1}))$

for some $A_{I_j} \in Alg(IMP_j)$ and $A_{E_{cj}} \in Alg(EXP_{cj})$.

Theorem 3 (Induced inheritance). If $C1$ `Wact` $C2$ via f, then

$$ACT(C1, f, C2) \texttt{ Wspec } C1 .$$

If $C1$ `Sact` $C2$ via f, then there exists an algebra $C3_{impl} \in Alg(BOD_3)$ such that

$$(ACT(C1, f, C2), C3_{impl}) \texttt{ Sspec } C1 .$$

Meyer (1986) showed how inheritance can simulate the genericity, by allowing generic classes in each object-oriented language; our model allows to specify some of the properties of the generic type parameters, while in most of the languages only signature can be specified.

Example 5. OBJ3 allows the specification of behavior through equational logics and the parameter properties are formalized in a theory. The actualization of a parametric class consists of providing a view, which represents a signature morphism, between the theory and the actual parameter. If the properties of the theory (parameter part) are derivable from those of the formal parameter up to the renaming effected by the provided view, the actualization results in the pushout object between the theory and the actual parameter.

The other relation that we are going to introduce relates a class $C2$, viewed as *producer* of its class interface, with another class $C1$, viewed as *consumer* of its import interface. Again we distinguish between a potential producer (weak notion) and a factual producer (strong notion).

Definition 8 (Combinable). Given classes $C1 = (C1_{spec}, C1_{impl})$ and $C2 = (C2_{spec}, C2_{impl})$

1. $C1$ is weakly combinable with $C2$, denoted by $C1$ `Wcomb` $C2$, if there exists a specification morphism h: $IMP_1 \rightarrow EXP_{c2}$;
2. $C1$ is strongly combinable with $C2$, denoted by $C1$ `Scomb` $C2$, if $C1$ `Wcomb` $C2$ via h and $V_h(V_{v_2}(C2_{\text{impl}})) = V_s(C1_{\text{impl}})$.

Interpretation. The relation `Wcomb` indicates that $C2$ can provide, through its class interface, the data and operations needed in the import interface of $C1$. The strong counterpart `Scomb` indicates that the specific realization of the class $C2$ provides exactly the import part of the chosen realization of $C1$.

The availability of a class $C2$ which can be combined to a class $C1$ determines an interconnection of classes equivalent to a single class which we are going to define now. The construction is similar to the composition of module specifications in Blum et al. (1987).

Definition 9 (Import passing). Let $Ci = (Ci_{\text{spec}}, Ci_{\text{impl}})$, $i = 1, 2$, be classes with $C1$ `Wcomb` $C2$ via h: $IMP_1 \rightarrow EXP_{c2}$. The combination of $C1$ and $C2$, denoted by $COMB(C1, h, C2)$, is the class specification $C3_{\text{spec}}$ with

- EXP_{i3} and EXP_{c3} are just EXP_{i1} and EXP_{c1}, respectively;
- IMP_3 is just IMP_2;
- the new implementation part is the union (pushout) of BOD_1 and BOD_2 with respect to IMP_1, with distinguished sort the one inherited from BOD_1;
- the new parameter part PAR_3 is the intersection (pullback) of PAR_1 and PAR_2 in (with respect to) EXP_{c2}.

The specification morphisms are the appropriate compositions deducible from the construction.

Interpretation. The new class specification $COMB(C1, h, C2)$ is obtained by *replacing* the import interface of $C1$ with the *product* of $C2$. The new implementation part is that of $C1$ where the yet-unimplemented part IMP_1 is replaced by the implemented part of EXP_{c2}, via the *fitting morphism h*. The specification IMP_1 is no longer the import, having been provided by EXP_{c2}, which in turn *needs* IMP_2, which becomes the overall import interface.

Theorem 4 (Induced semantics). The semantics of the class specification $COMB(C1, h, C2)$ is the set of all pairs $(A_{I_3}, A_{E_{c3}})$ such that

- $(A_{I_3}, A_{E_{c2}}) \in SEM(C2_{\text{spec}})$
- $(A_{I_1}, A_{E_{c3}}) \in SEM(C1_{\text{spec}})$
- $A_{I_1} = V_h(A_{E_{c2}})$

for some $A_{E_{c2}} \in Alg(EXP_{c2})$ and $A_{I_1} \in Alg(IMP_1)$.

The next theorem shows that combination is one way to obtain a new class by reusing an old one. This reuse is *predictable* since the semantics of the new class can be expressed using those of the combining classes.

Theorem 5 (Induced inheritance). If $C1$ `Wcomb` $C2$ via h, then

$$COMB(C1, h, C2) \texttt{ Wspec } C1 .$$

If, in addition, $s'_S(pt(BOD_2)) = pt(BOD_3)$ for s': $BOD_2 \to BOD_3$, then

$$COMB(C1, h, C2) \texttt{ Wreuse } C2 .$$

If $C1$ `Scomb` $C2$ via h, then there exists an algebra $C3_{impl} \in Alg(BOD_3)$ such that

1. $(COMB(C1, h, C2), C3_{impl})$ is a class (still denoted by $COMB(C1, h, C2)$);
2. $COMB(C1, h, C2)$ `Sspec` $C1$;
3. If, in addition, $s'_S(pt(BOD_2)) = pt(BOD_3)$ then $COMB(C1, h, C2)$ `Sreuse` $C2$.

There are several connections among these two relations and the interconnection mechanisms. It is also possible to extend the notion of Combination and Actualization to the case where one class provides only part of what another class needs. We can define in these cases notions of partial actualization and combination and obtain results similar to the induced inheritance mentioned above.

Most of the current languages allow to import from separately written classes. But they do not provide any combination mechanism since the importing procedure requires the name of the class from which we wish to import and produces an implicit combination of the imported code with what is being developed. The only way to require something, and then provide a supplier for it, is by means of the genericity and the actualization mechanism. Of course, this does not represent a satisfactory solution because it is just a special case of combination.

6 Concluding remarks

In this work, we have introduced a class model used for analyzing the class notion as defined in a number of object-oriented languages. Not all the features we consider have been found in the languages, e.g., encapsulation has not been treated uniformly, although it can be considered a basic ingredient of the object-oriented paradigm. Our model introduces also a formal import interface which is not present in the analyzed languages. In fact, as a rule the importing is realized through the so-called *use clause* which lists the names of already designed and implemented classes which are meant to be imported. The formal import provides a means to draw a boundary which allows us to define the semantics of a class in isolation, without relying on the semantics of the imported classes.

The proposed setting includes a rigorous treatment of the inheritance which has been classified in specialization inheritance and reusing inheritance. The distinction has been dictated by the necessity of avoiding confusion between the reuse of code without any constraint and the functional specialization. An important result shows how the specialization inheritance can be used in order to simulate the reusing one.

Other two mechanisms has been proposed: the parameter-passing and the import-passing. The former is used in order to replace the formal parameter by an actual one. Analogously, the import passing serves for replacing the formal import interface by the features exported by another class. Both techniques can be used for arranging classes in hierarchies. These hierarchies are closely related to inheritance hierarchies, as shown in Sect. 5: the producer C of an interface which satisfies the constraints of the parameter or of the import part of a class C' determines a new class which inherits from it by reusing and from C' by specialization. We believe this to be very important since it provides a restricted reuse of code where we can predict the behavior of the outcome.

In the framework presented here, the simplest form of algebraic specifications has been used, but both the formalizations and the results can immediately be extended to other frameworks based on institutions (Goguen and Burstall 1983) other than the equational one and on different specification logics (Ehrig et al. 1991). In particular, we plan to consider different semantics for the classes, such as the behavioral one.

We have outlined how the inheritance can be used at different levels, in particular the weak specialization can be used in order to complete requirement specifications in a monotonic or incremental way. Likewise, the weak reusing inheritance can be viewed as a non-monotonic design step which can be reduced to a monotonic one via the theorem which states how to simulate the reusing inheritance. On the other hand, some problems arise when the intended use of specialization degenerates in reusing inheritance. In current object-oriented languages it is not possible to refine a method without redefinition: a shadowing method can violate the invariants of the superclass because it does not assure the compatibility of related semantics. Refinement, as realized in current object-oriented programming, can therefore be considered harmful.

Different techniques can be used in order to enhance the quality of software. Among these, inheritance and genericity play an important role. An informal comparison between these two techniques can be found in Meyer (1986) where the author claims that inheritance is more powerful than genericity. Section 5 provides a formal proof of this result through Theorem 3.

In general, the model provides a quite extended support to the specification and design of classes but does not mention any notion of object. A uniform treatment of objects to be included in this framework is under investigation.

Acknowledgements

This research has been supported in part by CNR under the project "Sistemi Informatici e Calcolo Parallelo", subproject "Linguaggi di Nuova Concezione", by the European Communities under ESPRIT Basic Research Working Group COMPASS, and by the Deutsche Forschungsgemeinschaft.

References

Alencar, A. J., Goguen, J. A. (1991): Ooze: An object-oriented z environment. In: America, P. (ed.): ECOOP '91 European Conference on Object-Oriented Programming. Springer, Berlin Heidelberg New York Tokyo, pp. 180–199 (Lecture notes in computer science, vol. 512).

America, P. H. M. (1990): Designing an object-oriented programming language with behavioral subtyping. In: de Bakker, J. W., de Roever, W. P., Rozenberg, G. (eds.): Foundations of object-oriented languages. Springer, Berlin Heidelberg New York Tokyo, pp. 60–90 (Lecture notes in computer science, vol. 489).

Blum, E., Ehrig, H., Parisi-Presicce, F. (1987): Algebraic specifications of modules and their basic interconnections. J. Comput. Syst. Sci. 34: 293–339.

Ehrig, H., Mahr, B. (1990): Fundamentals of algebraic specification 2: module specifications and constraints. Springer, Berlin Heidelberg New York Tokyo (EATCS monographs in theoretical computer science, vol. 21).

Ehrig, H., Mahr, B. (1985): Fundamentals of algebraic specification 1: equations and initial semantics. Springer, Berlin Heidelberg New York Tokyo (EATCS monographs in theoretical computer science, vol. 6).

Ehrig, H., Baldamus, M., Orejas, F. (1991): New concepts of amalgamation and extension for a general theory of specifications. In: Bidoit, M., Choppy, C. (eds.): Recent trends in data type specification. Springer, Berlin Heidelberg New York Tokyo, pp. 199–221 (Lecture notes in computer science, vol. 655).

Goguen, J. A., Burstall, R. (1983): Introducing institutions. In: Clarke, E., Kozen, D. (eds.): Logics of programming. Springer, Berlin Heidelberg New York Tokyo, pp. 221–256 (Lecture notes in computer science, vol. 164).

Lehrmann Madsen, O., Magnusson, B., Moller-Pedersen, B. (1990): Strong typing of object oriented languages revised. In: Meyrowitz, N. (ed.): Proceedings Joint Conferences OOPSLA and ECOOP, Ottawa, Oct. 21–25, 1990. Association for Computing Machinery, New York, pp. 140–149.

Meyer, B. (1986): Genericity versus inheritance. In: Meyrowitz, N. (ed.): Proceedings ACM Conference on Object-Oriented Programming Systems, Languages and Applications, OOPSLA '86, Portland, Oregon, Sept. 29–Oct. 2, 1986. Association for Computing Machinery, New York, pp. 391–405.

Parisi-Presicce, F., Pierantonio, A. (1991): An algebraic view of inheritance and subtyping in object oriented programming. In: van Lamsweerde, A., Fugetta, A. (eds.): ESEC '91. Springer, Berlin Heidelberg New York Tokyo, pp. 364–379 (Lecture notes in computer science, vol. 550).

Schaffer, C., Cooper, T., Bullis, B., Killian, M., Wilpolt, C. (1986): An introduction to trellis/owl. In: Meyrowitz, N. (ed.): Proceedings ACM Conference on Object-Oriented Programming Systems, Languages and Applications, OOPSLA '86, Portland, Oregon, Sept. 29–Oct. 2, 1986. Association for Computing Machinery, New York, pp. 9–16.

Snyder, A. (1986): Encapsulation and inheritance in object-oriented programming languages. In: Meyrowitz, N. (ed.): Proceedings ACM Conference on Object-Oriented Programming Systems, Languages and Applications, OOPSLA '86, Portland, Oregon, Sept. 29–Oct. 2, 1986. Association for Computing Machinery, New York, pp. 38–45.

On subtyping in languages for symbolic computation systems

P. Di Blasio and M. Temperini

1 Introduction

The development of new software systems for symbolic computation, and for computer algebra in particular, has surged ahead in recent years. New systems have been developed from scratch, while others have been renewed and increased in power (Wang and Pavelle 1985, Hearn 1987, Char et al. 1991, Jenks and Sutor 1992, Buchberger et al. 1993). The new generation of symbolic computation systems was developed keeping in mind the evolution of powerful interfaces, allowing for the application of stated algorithms over established structures (Char et al. 1991, Wolfram 1991). However, symbolic computation and, in particular, computer algebra present particular necessities that need to be accomplished from a software engineering point of view. In some of these systems, the use of built-in programming languages has been considered as a tool for expressing new data structures and algorithms. Moreover, a large common software basis can be exploited as a library, in both developing algorithms and data structures, and enriching the library itself introducing new pieces of software. At present, the idea of constructing and maintaining libraries of mathematical software is being taken into consideration in the development of the SACLIB system (Buchberger et al. 1993), which is still in progress.

However, one aspect in the development of new systems apparently left out as yet is the adoption of a powerful software development methodology. It should be supported by a system language, in order to flank the specification of data structures and of the related algorithms. Leaving out this aspect is responsible for the complex maintenance needed by systems and, thus, for the difficulties found in the expansion of systems by reusing or modifying what is already defined in them. In fact, these difficulties increase their importance with the dimension of the whole system. Even if the existing code is organized in library items, that methodological deficiency may result in serious difficulties in establishing the intercommunication among the library modules used in a program. Think, for example, of the relationships that should hold – and, in some way, be tested – among data structures defined separately and then collected in the same algorithm.

In order to give a solution to mentioned problems, different aspects of programming languages have been investigated. Some researches used concepts of object-oriented programming (OOP) paradigm (Limongelli et al. 1992). The main needs that have been detected are the use of an *abstract data type* (ADT) language, the *viewing* feature, and the *hierarchical* organization of data structures and algorithms. The possibility of splitting the specification and implementation hierarchies has been recognized as a means for exploiting some general peculiarities, which are common to sets of different structures (Baumgartner and Stansifer 1990, Limongelli and Temperini 1992).

Axiom (Jenks and Sutor 1992) implements the concept of category and domain to express the duality of the specification and the implementation of the structures. Views (Abdali et al. 1986) features the viewing operation to enhance the expressive power of hierarchies: by viewing, it is possible to consider a concrete structure as instance of several abstract types, as is needed. In Limongelli and Temperini (1992) the data structures to be classified are distinguished into abstract, parametric, and ground. They are implemented by means of the respective class constructs of an OOP language. The hierarchies are imagined lying on three planes, some of them connected by the inheritance relationships existing between their structures.

In Temperini (1992), features of OOP methodology have been studied and selected in order to define a suitable mechanism of inheritance, supported by a language, for the design and implementation of symbolic computation systems. The requirements for polymorphism and redefinition of attributes in subclassing have been shown to be source of potential problems, mainly because of the possibility of incorrectness at run-time of statements that were checked and declared correct at compile-time. To be clear about what is meant by the term *correctness*: a program is correct if its statements which satisfy the rules of the compile-time type checking, never result in run-time type errors, due to the polymorphic behavior of some of their elements.

Summing up we want to define a strongly typed OOP language suitable as the software development tool of a symbolic computation system, which provides class structure to manage ADTs and supports multiple inheritance to model specialization hierarchies.

In this paper we provide the theoretical background for such a task. Based on the type system we are going to define, Di Blasio et al. (1997) present the inheritance mechanism and its implementation aspects.

A model of inheritance was first approached by Cardelli (1984), where method redefinition is constrained by a contravariant (or anti-monotonic) subtyping rule. This approach leads to a well-founded type theory in which the correctness problem is solved at the cost of redefining the method by the unnatural contravariant rule (Danforth and Tomlinson 1988).

On the contrary, a covariant redefinition rule is more natural and expressive from a software engineering point of view. Such a redefinition rule can be safely supported by a programming language, provided that a suitable mechanism of dynamic binding is defined in an environment of strict inheritance with redefinition (Regio and Temperini 1990).

The criticism to the contravariant redefinition rule has led to the definition of a type system allowing for covariant redefinitions in a multimethod object-oriented language (Agrawal et al. 1991). In Castagna et al. (1992), the contravariant subtyping rule is preserved, over arrow types, while a covariant one, for the treatment of overloaded functions, is presented. However, both these approaches present some restrictions in the construction of multiple inheritance hierarchies to guarantee the correctness of the type system in case of name conflict.

Starting from the principle of substitutability (Wegner and Zdonic 1988), that underlies any definition of subtyping, in this paper we present a type system for object-oriented languages based on a covariant subtyping rule. In this system all restrictions defined in the related approaches cited above, to force soundness of their type system, are relaxed.

The paper is organized as follows. In Sect. 2 we give an informal presentation of an enhanced mechanism of strict inheritance (*ESI*), motivating the main choices through a discussion on general aspects of programming on specialization hierarchies. In particular, we discuss *attribute redefinition* in subclassing, and *abstraction level* in the interpretation of language expressions. In Sect. 3 the covariant subtyping rule $\leq$ is presented. Type correctness of language expressions is proven. Section 4 concludes the paper with some further discussion about related works.

2 Motivations

To motivate the definition of our type system, we discuss an example concerning a well known problem of noncorrespondence between static and dynamic type-correctness of method invocations (Bruce 1993, Di Blasio and Temperini 1993), arising from the use of a covariant redefinition rule between methods.

Essentially, the problem arises from the identification of the subtyping and subclassing notions, which has been argued as incorrect by Cook et al. (1990).

The example we are discussing is given right to show how this coupling can be dangerous in actual languages.

In Fig. 1 a hierarchy composed of three classes is shown: `polygon`, `rectangle`, and `window`. We consider a window as a rectangle whose edges are parallel to the axes. The three classes are in subclassing relation. Let us suppose that there is also a subtyping relation among the types corresponding to the described classes.

We want to compute some intersections between objects of these three classes. So we define an intersection method in `polygon` which takes a parameter of class `polygon` and returns an instance of `polygon`. The operation of intersection between rectangles is obtained by specializing the inherited `Intersection` method in the class `rectangle`. The redefined method takes a rectangle as parameter and returns an object of class `polygon`. A similar redefinition takes place in the `window` class, to provide a method for intersections between windows. This is a natural way of modeling such a set of geometric figures, but the execution of the instructions included in Fig. 1 creates serious problems.

After some variables have been declared, two objects w_1 and r_1 of class `window` and `rectangle`, respectively, are instantiated using the `Create` instruction. Then the object referred to by w_1 is polymorphically assigned to the variable p_1, being `window` subclass of `polygon`. Let us note that the method invocation p_1`.Intersection(`r_1`)` is statically correct because the intersection

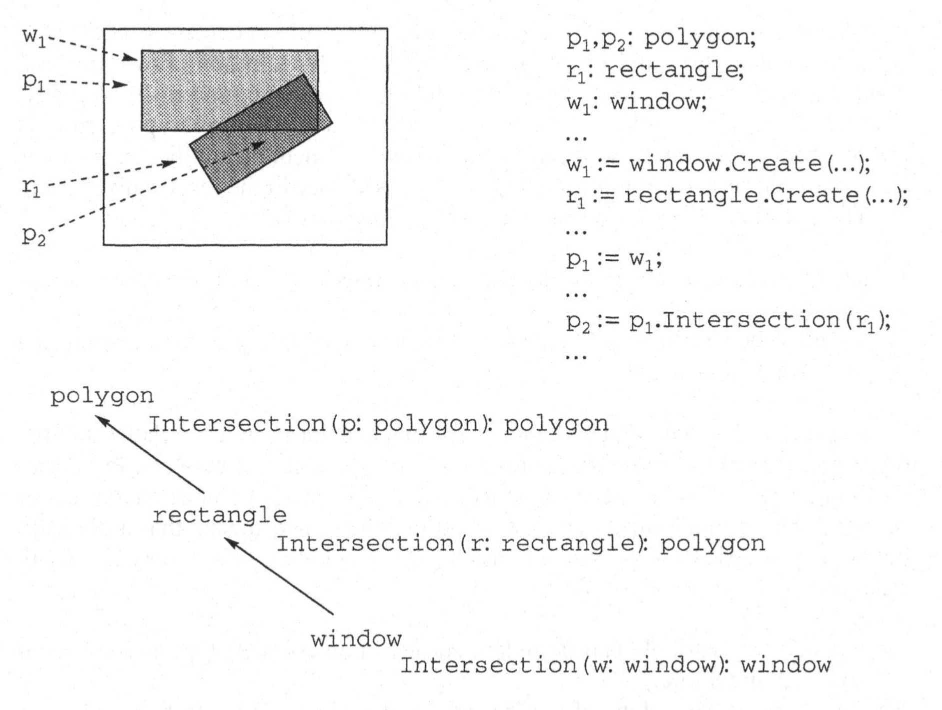

Fig. 1. Unsafeness of a syntactically correct method invocation

method defined in the class of p_1 is applicable to an object r_1 of class `rectangle`.

At run-time, as established by the usual dynamic binding rules, the code version of `Intersection` method is selected in the class `window`, since p_1, after the execution of the polymorphic assignment, refers to an object of `window` class. But the execution is incorrect because the actual parameter r_1 should be of class `window`, while it is actually of class `rectangle`.

Thus a type failure occurs, due to an unsatisfactory treatment of the coexistence of polymorphism and method redefinition.

This example creates troubles for most well known statically type-checked object-oriented languages, such as Eiffel (Meyer 1992), C++ (Stroustrup 1991), Modula-3 (Cardelli et al. 1988). In the original version of Eiffel this example gave a run-time type failure. The Eiffel version 3.0 works around this problem, giving eventually an unsatisfactory solution (Bruce 1993). C++ and Modula-3 avoid this problem by not allowing the user for redefining covariantly the types of method parameters in subclasses.

The key question is how the methods can be redefined and their types can be changed in subclassing, to manage polymorphism correctly. A first solution is to adopt the well known contravariant subtyping rule (Cardelli 1984). When this solution is chosen, no subtyping relation can be established among the classes of the previous example. So it is possible to have only a subclassing relation among the given classes, and polymorphic uses of their instances are not legal.

A second solution is to adopt a covariant subtyping rule between methods as in Agrawal et al. (1991) and Castagna et al. (1992). This solution allows for both defining specialization class hierarchies and managing polymorphism correctly.

Our approach involves the introduction of a covariant subtyping rule. It differs from those latter solutions in the way the method conflict problem is treated for hierarchies of multiple inheritance, as we see in the rest of this section.

The solution we propose for this problem consists of:

1. an inheritance mechanism based on strict inheritance with covariant redefinition of attributes;
2. a static type checking of language instructions; in particular, we are interested in method invocation.

Regarding (1), we want an inheritance mechanism in which subclassing the relationship between classes implies subtyping the relation between the corresponding types. Class structure contains a set of instance variables and a set of methods which make up its attributes. Strict inheritance means that a class inherits all the attributes of its superclasses possibly adding other ones. Covariant redefinition means that:

1.a. an instance variable can be redefined in a subclass if its type is subtype of the redefined one;

1.b. method can be redefined if types of its arguments and result are subtypes of the corresponding arguments and result types of the redefined one.

Regarding (2), to guarantee the correctness of a given method invocation, the static type-checking must ensure the following requirements:

2.a. for such method invocation, an executable method has to exist at run-time;

2.b. the type of the result of the method execution, has to be subtype of the one inferred for the method invocation at compile-time.

These requirements allow us to establish the correspondence between static and dynamic correctness of type checked method invocations.

Condition (2.a) is simply fulfilled by the existence, within the class hierarchy, of a method which is applicable to the static invocation. This is enough since any variable ranges at run-time on objects which are subtype of its statically declared type.

Let us note that in case of single inheritance condition (2.a) encompasses condition (2.b). This is no more true in case of multiple inheritance, as the example of Fig. 2 shows: methods `m` in classes `A` and `B` return incompatible types $R_{\mathtt{A}}$ and $R_{\mathtt{B}}$; `m` is not redefined into `C`. For the invocation a.`m`(b), type $R_{\mathtt{B}}$, for instance, is inferred statically. The only existence of the method `m` in `B` does not guarantee that the result at run-time will be subtype of $R_{\mathtt{B}}$, because both methods are applicable so we could apply also method `m` defined in `A`. What method `m` should be chosen at compile- and run-time? The solution to such dilemma is called *conflict resolution*.

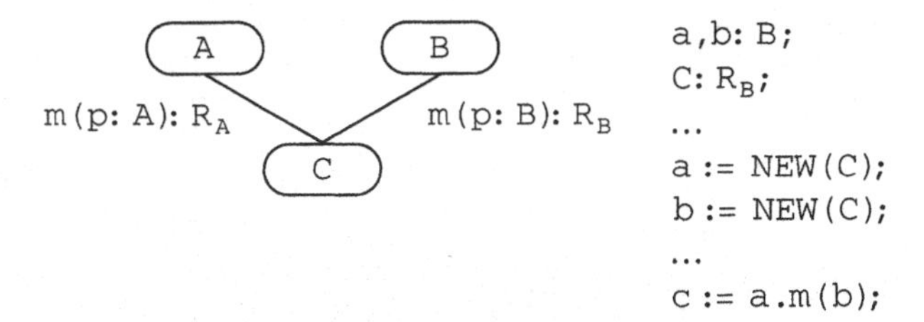

Fig. 2. Problem of conflict resolution

This problem has various criteria to be met. Different solutions have been proposed in literature to such purpose. We can distinguish such solutions into two families, with regard to conflict resolution requesting an interaction with the user or not. Still looking at Fig. 2, in the first case the user is requested either to select explicitly the method to be executed just in the invocation (like in C++), or to redefine the method `m` into class `C` (like in Castagna et al. 1992) or to rename one of the inherited `m`. In the second case the conflict can be solved just applying linearization techniques. On multiple inheritance systems, either a *global type order* or a *local type order* could be applied [CLOS (Bobrow et al. 1988) is an example of language using local linearization technique]. Ordering on types involves some ordering on methods too. Therefore this approach needs some constraint on result types of confusable methods, to ensure condition (2.b) on compatibility of result types (Agrawal et al. 1991, Ghelli 1991). For instance, if `A` precedes `B`, we can establish an ordering between the two confusable methods: `m`: $\mathtt{A} \rightarrow R_{\mathtt{A}}$ precedes `m`: $\mathtt{B} \rightarrow R_{\mathtt{B}}$. So, the constraint on result types is $R_{\mathtt{A}} \leq R_{\mathtt{B}}$. Under the above conditions the execution of the invocation `a.m(b)` in Fig. 2 results in the application of the method `m` in `A` which returns a result of type $R_{\mathtt{A}}$ subtype of $R_{\mathtt{B}}$ (the type inferred at compile-time).

Let us look at an example of method name conflict (Fig. 3). We want to inherit both the methods `mission`, guaranteeing result type compatibility at the same time. Let us examine how the approaches cited above behave on this example.

Result-types of methods `mission` in `doctor-student` and `fellow-researcher` classes are not compatible, so the solution of linearization techniques cannot be accomplished, also because we want to inherit both methods. We could redefine method `mission` into `doctor&fellow`; but no redefinition makes sense in the example, since the `mission` behavior should be redefined just by choosing one of the two methods in superclasses. Also renaming one of methods `mission` is possible, however, we want a solution such that the user remains free also from renaming requirement.

Moreover, let us consider the program statements listed in Fig. 3. Object `df` of class `doctor&fellow` could be asked to execute method `mission` in two different *abstraction* contexts: if taken from array `a`, it is a particular doctor-student; if taken from array `b`, it is a particular fellow-researcher. In case (a), `df` is a doctor&fellow to be "considered" as a doctor-student. Hence, method `mission` of class `doctor-student` must be executed. On the other hand, if `df`

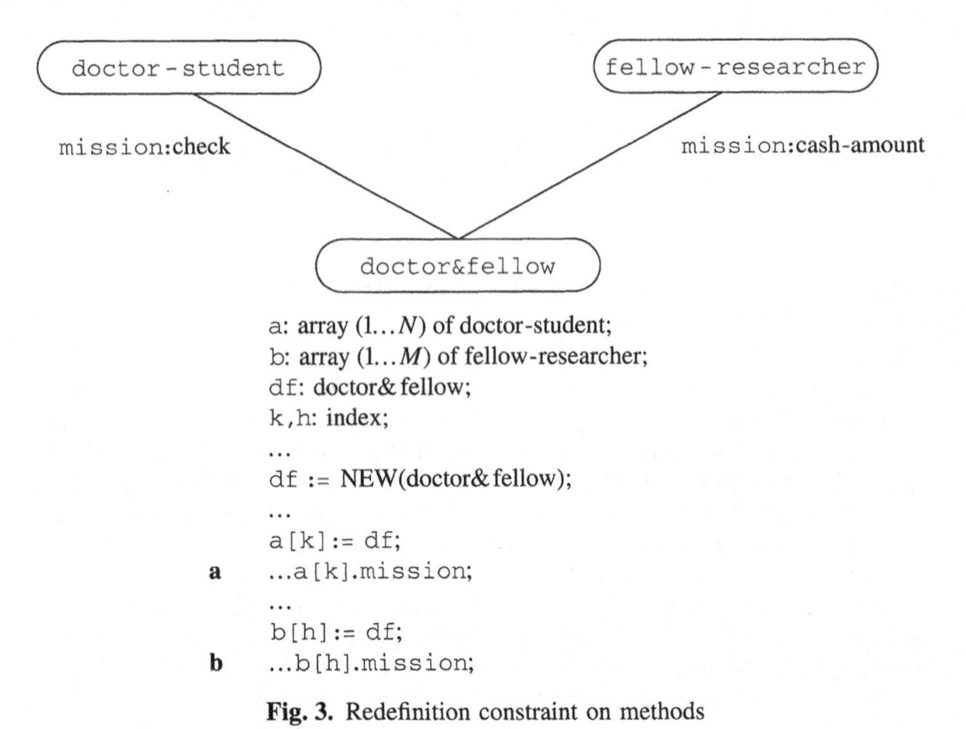

Fig. 3. Redefinition constraint on methods

is taken from the collection of fellow-researcher, like in case (b), it is a doctor&fellow to be "considered" as a fellow-researcher. Hence, method `mission` of class `fellow-researcher` must be executed.

Let us note that the known conflict resolution criteria neither are applicable when method redefinition is not reasonable or result types of confusable methods are not compatible, nor they are able to take into consideration the *abstraction level* established by the program for certain polymorphic objects.

So it makes sense to extend the specialization inheritance, such that the cited constraints are relaxed and a suitable mechanism for taking into consideration the abstraction level is provided.

What we need is a method lookup based on a searching procedure working on a subgraph of the inheritance one, bounded by the run-time class of the object to which the message was sent and the compile-time type of the expression which refers to that object.

If this approach is followed, we can exploit the fact that variable `a[k]`, in Fig. 3, was declared of class `doctor-student`, so `mission` behavior, bound to the run-time object, is taken from `doctor-student`. The method lookup algorithm is forced to look for an executable method only in the subgraph composed by classes `doctor-student` and `doctor&fellow`, excluding the `fellow-researcher` class. A symmetric behavior is followed for the method invocation of case (b).

Summing up, the features of our proposal are as follows: it eliminates the previously needed constraints and enables a more flexible inheritance mechanism; it defines a method lookup sensitive to abstraction level (i.e., class that was decided at compile-time for the object expression which the message is sent to); it guarantees the compatibility of result type of a method invocation between compile-time and run-time, as we prove in the next section.

3 A model of subtyping

The most generally accepted idea of subtyping is the following principle of substitutability (Wegner and Zdonic 1988): given two types S and T, we say that S is subtype of T if it is possible to have an object of type S whenever an object of type T is expected. The aim of this section is to provide the definition of a subtyping rule based on covariant relation between functions, such that it is able to model this idea of subtyping.

The subtyping relation we are going to define is stated over the following type structure. Such a type structure is denoted in terms of its typed data attributes (VARIABLES) and functional attributes (FUNCTIONS).

TYPE C:
VARIABLES $v_1\colon T_1; \ldots; v_m\colon T_m$
FUNCTIONS $m_1\colon P_1 \to R_1; \ldots; m_n\colon P_n \to R_n$

The declaration $v_i\colon T_i$ denotes a variable v_i of type T_i, while $m_i\colon P_i \to R_i$ denotes a function which takes an argument of type P_i and returns a result of type R_i.

We adopt such a structure in order to state the distinction between the structural and functional component of a type. Any concrete instance of type C is also called an *object of type* C. Since each function is defined in a stated type specification, the arguments of its signature do not indicate the type in which the function itself is defined. In order to make our notations during the following discussion simpler, without any loss of generality, the functional attributes have only one argument.

Now we can define the covariant subtyping rule $\leq$.

Definition 1 (Subtyping rule $\leq$).
1. Given type C, C is a subtype of C ($C \leq C$);
2. let C and C' be:
   type C:
   VARIABLES $v_1\colon T_1; \ldots; v_m\colon T_m$
   FUNCTIONS $m_1\colon P_1 \to R_1; \ldots; m_n\colon P_n \to R_n$
   type C':
   VARIABLES $v_1\colon T'_1; \ldots; v_m\colon T'_m; \ldots; v_{m+q}\colon T'_{m+q}$
   FUNCTIONS $m'_1\colon P'_1 \to R'_1; \ldots; m'_p\colon P'_p \to R'_p$
   we say that C' is a *subtype* of C ($C' \leq C$) iff the following conditions do hold:
   a. $T'_1 \leq T_1; \ldots; T'_m \leq T_m$;

b. if $m_i = m'_j$ where $i \in \{1 \ldots n\}$ and $j \in \{1 \ldots p\}$ then $P'_j \leq P_i$, and $R'_j \leq R_i$;
c. $C' \leq \bar{C}$ for all $\bar{C}$ such that $C \leq \bar{C}$.

Note that $m_i = m'_j$ means that the same identifier is used for the functions m_i and m'_j.

Given $C' \leq C$, the type C can also be called a *supertype* of C'. We can also use $C' < C$ to mean *strict subtyping*, i.e., ($C' \leq C$ and $C' \neq C$).

A graphical interpretation of a set of types defined according to the subtyping rule $\leq$ is a direct acyclic graph (DAG), in which each arc starts from a supertype and leads to a subtype. So we can speak in terms of subtyping hierarchy.

Considering the syntax of the language expressions, in the following the letters e, f, g, ... stand for language expressions, o, p, q, ... are object identifiers, and C, O, P, Q, ... are types.

Language expressions have the following form:

$$e ::= o^V \mid e.v \mid e.m(e)$$

where o is an object identifier, whose index represents the type which it belongs to, $e.v$ and $e.m(e)$ stand respectively for the access to variable attribute v in the object returned by the expression e and for the invocation of function m on the object returned by the expression e, with argument e.

We omit indexing object identifiers explicitly, when their types are either clear from context or meaningless.

It is simple to see that variable access and function invocation make sense (we say they are *legal*) only if certain conditions do hold. For instance, $o^T.v$ is legal only if a variable v is defined in T. Moreover, $o.m(q^Q)$, is legal only if a function m is defined in T, or in some supertype of T, such that the type Q is subtype of the type of the formal parameter declared for m. Such a function m is said to be *applicable* to o and q. Note that we use the subsumption rule implicitly. This rule states that o^T is also an object of any supertype of T.

Before formally defining type correct language expressions, by giving the set of type checking rules, let us introduce some notations useful in the following.

We say that an attribute defined in a type C is *redefined* in a subtype C', if it is defined in both types, and in them it either has different type in case of data attributes, or different domain and/or codomain in case of functional attributes. We say that the attribute definition in C' *covers* the one in C.

Definition 2 (Attribute covering). Given the types C and C', with $C' \leq C$, we say that

1. the variable v_j: T'_j in C' covers the variable v_j: T_j in C iff $T'_j < T_j$;
2. the function m'_j: $P'_j \rightarrow R'_j$ in C' *covers* the function m_i: $P_i \rightarrow R_i$ in C, with $m'_j = m_i$, iff either $P'_j < P_i$ or $R'_j < R_i$ or both.

We can assert that a covering attribute is a specialization of the covered one.

With respect to Definition 2(2), we also note that since a function m, defined into the type C, could be covered by different redefinitions in subtypes of C, it follows that starting from any legal function invocation, several different applicable functions could exist in different but related types. Among such applicable functions, one has to be applied as much specialized as possible.

Below we set up the criterion for determining such a *selected function*. Given a function invocation $e.m(e)$, we identify as $App_{e.m(e)}$ the set of all types in which a function m is defined such that the invocation is legal. Note that $App_{e.m(e)}$ can be empty. In this case, of course, the invocation is illegal. For instance, if we have the invocation $o^O.m(q^Q)$, $App_{o.m(q)}$ is composed of all supertypes of O in which a function m is defined applicable to q^Q.

It is simple to note that, once a set of types has been defined following the subtyping rule of Definition 1, the resulting hierarchy is a partially ordered set (poset) under the subtyping relation $\leq$. Obviously $App_{e.m(e)}$ is a subset of such a poset. This property should be carefully considered with respect to function invocations. Actually a legal invocation, say $o.m(q)$, could give rise to two dramatically different configurations for the related $App_{o.m(q)}$, depending on the existence or nonexistence of the *bottom* element B in it (i.e., an element $B \in App_{o.m(q)}$ such that $\forall B' \in App_{o.m(q)}$, $B \leq B'$):

1. if $App_{o.m(q)}$ has a *bottom* element B then the function m can be correctly selected from B itself; such m is, among the applicable functions, the most specialized;

2. if $App_{o.m(q)}$ admits more than one minimal element w.r.t. the $\leq$ ordering, then such elements contain several applicable and *equally most specialized* functions m. In fact, all minimal elements of $App_{o.m(q)}$ are types whose functional attributes m are equally applicable. Normally the problem of choosing one single m from all the applicable ones is solved by adding further selection criteria. In the following, $\mathcal{S}$ denotes the application of such a criterion.

Let $MIN(App_{o.m(q)})$ be the set of minimal elements of $App_{o.m(q)}$ w.r.t. $\leq$, then the selected function m is defined in one of the types of this set. In case (2) it should be chosen by the mentioned criterion $\mathcal{S}(MIN(App_{o.m(q)}))$. In case (1) *MIN* has only one element (the bottom of $App_{o.m(q)}$) so the selected function is uniquely determined.

Now using the definitions given above we define the *type checking rules* for the language expressions:

$$\vdash o^V: V\,, \tag{1}$$

$$\frac{\vdash e: R \quad v: V \in R}{e.v: V}\,, \tag{2}$$

$$\frac{\vdash e: O \quad \vdash e': Q \quad C = \mathcal{S}(MIN(App_{e.m(e')})) \quad m: P \to R \in C}{\vdash e.m(e'): R}\,. \tag{3}$$

Before proving that the rule of Definition 1 models the principle of substitutability, let us stress what means replacing an object o^O in a language expression e by an object $\bar{o}$ of type $\bar{O} \leq O$, eventually obtaining the language expression $\bar{e}$:

1. given a type correct variable access $e.v$, then $\bar{e}.v$ is type correct too, and its use doesn't violate the context of the $e.v$ occurrence;

2. given a type correct function invocation $e.m(e')$, with e': Q, then $\bar{e}.m(e')$ is type correct too, and its use doesn't violate the context of the function invocation occurrence.

We identify, here, the context of both a variable access and a function invocation with the type expected for the result of their execution.

For instance, in the context of an assignment – like $x := e.v$ or $x := e.m(e')$ – the expected type for $e.v$ and $e.m(e')$ has to be subtype of the type of x. In the context of a function invocation – like $e_1.m_1(e.v)$ or $e_1.m_1(e.m(e'))$ – the expected type for $e.v$ and $e.m(e')$ has to be a subtype of the type of the correspondent formal parameter of m_1.

The previous notes about the coherence with the context of an expression can be summarized by the following statements.

In case (1), the result type of $\bar{e}.v$ must be a subtype of the type expected for $e.v$.

In case (2), the result type of $\bar{e}.m(e')$ must be a subtype of the type expected for $e.m(e')$.

A replacement in a variable access, as in case (1), is always safe if $\bar{O} \leq O$, due to the covariant redefinition stated by the subtyping rule. The same safeness is not guaranteed in case (2): one must be more careful when searching for the function that answers an invocation, once a replacement has occurred. In this case, e.g., we have to think of the context in which such an invocation appears. The problem is the same we examined in Sect. 2, however, here we consider it in using the definition introduced above.

Let us take, for instance, the function invocation $o^O.m(q^Q)$. Once o is replaced by $\bar{o}$: $\bar{O} \leq O$ in the function invocation, at least the function m in $\mathcal{S}(MIN(App_{o.m(q)}))$ could be executed without violating the context. Actually this might not be the most specialized function which can be applied. Neither we can choose the function to be executed directly in $MIN(App_{\bar{o}.m(q)})$, since in

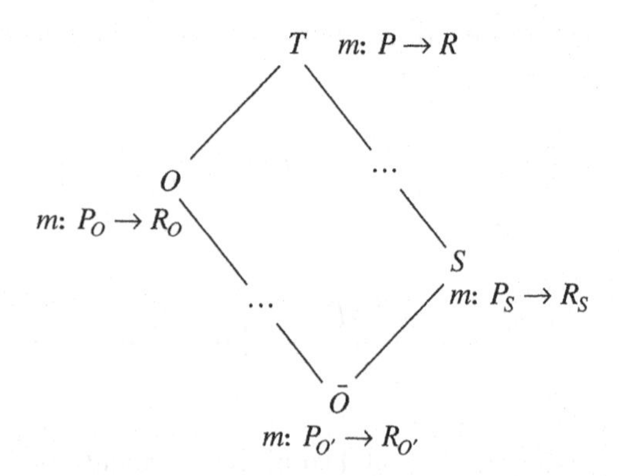

Fig. 4. Context of a function invocation

this minimal set, the case of Fig. 4 could occur, in which a function returns a result of the wrong type. In this example the function invocation $o^O.m(q^{P_O})$ is in a context of type at least R_O (we can suppose here, without any loss of generality, that $\mathcal{S}(MIN(App_{o.m(q)}))$ is exactly O). This context could be violated, after the replacement of o by $\bar{o}$, because of the possible call of the function m defined in the type S – suppose $S = \mathcal{S}(MIN(App_{\bar{o}.m(q)}))$, and $R_S \not\leq R_O$.

The following definition establishes the conditions to infer the type of a function invocation so that safe replacement can be managed also in this case.

Definition 3 (Inference rule (3′)). Let be $e_1.m(e_2)$: R a type correct function invocation, with e_1: O and e_2: Q, and $C = \mathcal{S}(MIN(App_{e_1.m(e_2)}))$. Moreover, let $\bar{e}_1.m(\bar{e}_2)$ be the expression obtained through a correct replacement in e_1 and/or e_2, with eventually $\bar{e}_1$: $\bar{O}$ and $\bar{e}_2 : \bar{Q}$. The type of $\bar{e}_1.m(\bar{e}_2)$ is inferred by the following rule:

$$\frac{\vdash \bar{e}_1\colon \bar{O} \quad \vdash \bar{e}_2\colon \bar{Q} \quad C' = \mathcal{S}(MIN(App_{\bar{e}_1.m(\bar{e}_2)} \cap \{S \mid S \leq C\})) \quad m\colon P' \to R' \in C'}{\vdash \bar{e}_1.m(\bar{e}_2)\colon R'}.$$

Now we can establish formally that Definition 1 models the concept of subtyping.

Theorem 1. Given a language expression e: R in which an object o^O occurs, if we replace o by an object $\bar{o}^{\bar{O}}$ ($\bar{O} \leq O$), then the resulting expression $\bar{e}$ is of type $R' \leq R$.

Proof. By induction on the structure of the language expressions.

$e \equiv o$: straightforward.

$e \equiv f.v$: we have f: O and v: R in O, so e: R by rule (2). Then we can replace o only in f, obtaining, by induction hypothesis, an expression $\bar{f}$ of type $\bar{O} \leq O$. By Definition 1, type $\bar{O}$ contains a variable v of type $R' \leq R$; hence the thesis comes from rule (2).

$e \equiv e_1.m(e_2)$: we have e_1: O and e_2: Q, $C = \mathcal{S}(MIN(App_{e_1.m(e_2)}))$ and a function m: $P \to R \in C$, so $e_1.m(e_2)$: R by rule (3). We can replace o in e_1 and e_2, obtaining, by induction hypothesis, the expressions $\bar{e}_1$: $\bar{O} \leq O$ and $\bar{e}_2$: $\bar{Q} \leq Q$. Now we can apply rule (3′), to infer $\bar{e}_1.m(\bar{e}_2)$: R'. Let us note that the existence of a type C' is guaranteed by the fact that at least the function m in C is applicable to $\bar{e}_1$ and $\bar{e}_2$. Moreover, type of $\bar{e}_1.m(\bar{e}_2)$, where m is the one defined in C', is a subtype of the type of $e_1.m(e_2)$, where m is the one defined in C ($C' \leq C$), due to the covariant subtyping rule stated on functional attributes. □

Concluding this section we examine the following example in order to show the main features of the subtyping model described above.

Let us assume that the following types have been defined:

TYPE *polygon*
VARIABLES ...
METHODS ... *Intersection*(p: *polygon*): *polygon* ...
TYPE *rectangle*
VARIABLES ...
METHODS ... *Intersection*(r: *rectangle*): *polygon* ...
TYPE *window*
VARIABLES ...
METHODS ... *Intersection*(w: *window*): *window* ...

having the subtyping relation $window \leq rectangle \leq polygon$.

Let $p_1.Intersection(r_1)$ be a function invocation with p_1: *polygon* and r_1: *rectangle*. Applying rule (3) we infer $p_1.Intersection(r_1)$: *polygon*.

Once p_1 is replaced by an object w_1 of type *window* we have:

1. $App_{w_1.Intersection(r_1)} = \{polygon, rectangle\}$,
   $window \notin App_{w_1.Intersection(r_1)}$,
   because the argument r_1 makes the function *Intersection* of *window* not applicable;
2. *rectangle* is the bottom element of $App_{w_1.Intersection(r_1)}$;

So, by choosing the most specialized applicable function *Intersection* in *rectangle* it is possible to replace an instance of the supertype *polygon* with one of the subtype *window*. Moreover, the compatibility of the result is preserved after the replacing, because the expected result of this legal invocation, on variable p_1, and the effective one, on object w_1, are objects of the same type *polygon*.

4 Conclusions

In this paper we presented a model of subtyping for an object-oriented programming language, suitable for the specification of data structures and algorithms of a symbolic computation system. The model is based on a covariant rule between type attributes. So a specialization relation among types is modeled and management of polymorphic instructions is made safe.

From this subtyping relation, a mechanism of attribute redefinition raises more natural w.r.t. those based on contravariant subtyping rule. This achievement is of particular relevance in languages for symbolic computation systems, where subtyping among abstract data structures is intended as specialization.

The correspondence between the static and dynamic correctness of programs is guaranteed by a set of rules for static type-checking of basic statements in which polymorphism is involved.

One interesting aspect of this model is the possibility of taking into account the abstraction level to which an object should be considered, while evaluating expressions in which it is involved. We showed examples in which such feature is meaningful.

Acknowledgement

This work has been partially supported by MURST and by CNR under the project "Sistemi Informatici e Calcolo Parallelo", grant no. 92.01604.69.

References

Abdali, S. K., Cherry, G. W., Soiffer, N. (1986): An object oriented approach to algebra system design. In: Char, B. W. (ed.): Proceedings ACM Symposium on Symbolic and Algebraic Computation, Symsac '86, July 21–23, 1986, Waterloo, Ontario. Association for Computing Machinery, New York, pp. 24–30.

Agrawal, R., DeMichiel, L. G., Lindsay, B. G. (1991): Static type checking of multimethods. In: Paepke, A. (ed.): Proceedings ACM Conference on Object-Oriented Programming Systems, Languages and Applications, OOPSLA '91, Phoenix, Arizona, Oct. 6–11, 1991. Association for Computing Machinery, New York, pp. 113–128.

Baumgartner, G., Stansifer, R. (1990): A proposal to study type systems for computer algebra. Tech. Rep. 90-87.0, RISC Linz, Johannes Kepler University, Linz.

Bobrow, D. G., DeMichiel, L. G., Gabriel, P., Keene, S. E., Kiczales, G., Moon, D. A. (1988): Common Lisp object system specification, X3JI3 doc. 88-002R, ANSI Common Lisp Standard Committee (also in SIGPLAN Not. 23/9).

Bruce, K. B. (1993): Safe type checking in a statically-typed object-oriented programming language. In: Proceedings Symposium on Principles of Programming Languages, Charleston, South Carolina, Jan. 10–13, 1993. Association for Computing Machinery, New York, pp. 285–298.

Buchberger, B., Collins, G. E., Encarnacion, M., Hong, H. (1993): A SACLIB 1.1 user's guide. Tech. Rep. 93-19, RISC Linz, Johannes Kepler University, Linz.

Cardelli, L. (1984): A semantics of multiple inheritance. In: Kahn, G., MacQueen, D., Plotkin, G. (eds.): Symposium on Semantics of Data Type. Springer, Berlin Heidelberg New York Tokyo, pp. 51–67 (Lecture notes in computer science, vol. 173).

Cardelli, L., Donahue, J., Galssman, L., Jordan, M., Kalsow, B., Nelson, G. (1988): Modula-3 report. Tech. Rep. SRC-31, DEC Systems Research Center, Palo Alto, CA.

Castagna, G., Ghelli, G., Longo, G. (1992): A calculus for overloaded functions with subtyping. In: Proceedings ACM Conference on LISP and Functional Programming, San Francisco, California, Jun. 22–24, 1992. Association for Computing Machinery, New York, pp. 182–192.

Char, B. W., Geddes, K. O., Gonnet, G. H., Leong, B. C., Monagan, M. B., Watt, S. M. (1991): Maple V language reference manual. Springer, Berlin Heidelberg New York Tokyo.

Cook, W., Hill, W., Canning, P. (1990): Inheritance is not subtyping. In: Hudak, P. (ed.): Proceedings Symposium on Principles of Programming Languages, San Francisco, California, Jan. 17–19, 1990. Association for Computing Machinery, New York, pp. 125–135.

Danforth, S., Tomlinson, C. (1988): Type theories and object-oriented programming. ACM Comput. Surv. 20: 29–72.

Di Blasio, P., Temperini, M. (1993): Subtyping inheritance in languages for symbolic computation systems. In: Miola, A. (ed.): Design and implementation of symbolic computation systems. Springer, Berlin Heidelberg New York Tokyo, pp. 107–121 (Lecture notes in computer science, vol. 722).

Di Blasio, P., Temperini, M., Terlizzi, P. (1997): Enhanced strict inheritance in TASSO-L. In: Miola, A., Temperini, M. (eds.): Advances in the design of symbolic computation systems. Springer, Wien New York, pp. 179–195 (this volume).

Ghelli, G. (1991): A static type system for message passing. In: Paepke, A. (ed.): Proceedings ACM Conference on Object-Oriented Programming Systems, Languages and Applications, OOPSLA '91, Phoenix, Arizona, Oct. 6–11, 1991. Association for Computing Machinery, New York, pp. 129–145.

Hearn, A. C. (1987): Reduce-3 user's manual. Rand Corporation, Mississauga, Ont.

Jenks, R. D., Sutor, R. S. (1992): Axiom, the scientific computation. Springer, Berlin Heidelberg New York Tokyo.

Limongelli, C., Miola, A., Temperini, M. (1992): Design and implementation of symbolic computation systems. In: Gaffney, P. W., Houstis, E. N. (eds.): Proceedings IFIP TC2/WG2.5 Working Conference on Programming Environments for High Level Scientific Problem Solving, Karlsruhe, Germany, Sept. 23–27, 1991. North-Holland, Amsterdam, pp. 217–226.

Limongelli, C., Temperini, M. (1992): Abstract specification of structures and methods in symbolic mathematical computation. Theor. Comput. Sci. 104: 89–107.

Meyer, B. (1992): Eiffel: the language, 2nd edn. Prentice Hall, Englewood Cliffs.

Regio, M., Temperini, M. (1990): Type redefinition and polymorphism in object-oriented languages. In: Proceedings 3rd Conference on Technology of Object-Oriented Languages and Systems, TOOLS PACIFIC, Darling Harbour, Sydney, Nov. 28–30, 1990, pp. 93–102.

Stroustrup, B. (1991): The C++ programming language. Prentice Hall, Englewood Cliffs.

Temperini, M. (1992): Design and implementation methodologies for symbolic computation systems. Ph.D. thesis, University "La Sapienza", Rome, Italy.

Wang, P., Pavelle, R. (1985): MACSYMA from F to G. J. Symb. Comput. 1: 69–100.

Wegner, P., Zdonic, S. (1988): Inheritance as an incremental modification mechanism or what like is and isn't like. In: Gjessing, S., Nygaard, K. (eds.): ECOOP '88 European Conference on Object-Oriented Programming. Springer, Berlin Heidelberg New York Tokyo, pp. 55–77 (Lecture notes in computer science, vol. 322).

Wolfram, S. (1991): Mathematica: a system for doing mathematics by computer. Addison-Wesley, Reading, MA.

Enhanced strict inheritance in TASSO-L

P. Di Blasio, M. Temperini, and P. Terlizzi

1 Introduction

In this paper we present the *ESI* mechanism and the related features of TASSO-L language. TASSO-L is the strongly typed object-oriented language embedded into TASSO environment (Miola 1997). It is devised to support the development of *ESI*-based object-oriented programs.

ESI is the inheritance mechanism derived by the simple type system presented in Di Blasio and Temperini (1997), so it is a subtyping inheritance based on a covariant subtyping rule. In describing *ESI* we refer to a small, yet significant, example dealing with different kinds of matrices ranging over integer elements. In this example we meet the problems arising in treatment of polymorphic instructions, when redefinitions of class features occur through subclassing. The subclassing rule is shown to be coherent with the cited subtyping model. The static type checking rules provided by the type system allow to establish the compile-time correctness of polymorphic statements. Their run-time correctness is preserved by a suitable method lookup algorithm. Once such static type checking is performed on programs, the amount of work needed by the run-time system decreases, making the execution much more efficient. This aspect should be of considerable importance, since very strict efficiency requirements arise in the field of symbolic computation.

The implementation of TASSO-L is built up as an extension of *Common Lisp Object System* (Bobrow et al. 1988). A relevant set of constructs have been either added or modified, by exploiting the *Meta-Object Protocol* provided by standard CLOS (Kiczales et al. 1991).

The paper is organized as follows. In Sect. 2 *ESI* is defined, to fit in the theoretical background of Di Blasio and Temperini (1997). In Sect. 3 the operational semantics of *ESI* is given via the method lookup algorithm. This is also the basis for the implementation of our inheritance mechanism in TASSO-L. In Sect. 4 the main characteristics of the programming system are presented. Syntax is explained, limited to constructs useful for examples. Section 5 provides some further discussion about related works and future developments.

2 Enhanced strict inheritance

First we consider the class structure on which *ESI* is defined.

Class structure contains a set of instance variables declared of a certain type and a set of methods. Each method is declared through its signature, which specifies types of formal parameter and result. Body of the method is also included in its declaration. In the following definition, *inheritance list* specifies the set of direct superclasses; v_j: C_j denotes an instance variable v_j of type C_j; m_i (p_i: P_i): R_i denotes a method signature with argument p_i of type P_i and a result type R_i.

Definition 1 (Class structure).
class *name* {*inheritance list*}
  v_1: C_1; ...; v_m: C_m;
  $m_1(p_1$: $P_1)$: R_1 is {m_1 *body*};
  ...
  $m_n(p_n$: $P_n)$: R_n is {m_n *body*};
endclass

We note that within the definition of a class we should distinguish between instance variables which are hidden or visible outside of class, and, therefore, between encapsulated class structure and interface. Class interface would contain only visible instance variables, besides methods. Moreover, the signature of a method should include the declaration of any number of parameters and of a result. However, in order to avoid a too cumbersome notation, without any loss of generality, we consider inheritable all instance variables defined in a class, and use methods having only one argument.

We denote with $\leq_{ESI}$ the subtyping inheritance relationship we are going to define.

ESI is equal to *strict inheritance* plus *covariant redefinition*. Strict inheritance means that a class inherits all attributes of its superclasses, possibly adding other ones.

An inherited attribute can be *redefined* in the subclass: this means that its previous definition is replaced by a new one, with some modifications. The redefinition is covariant if it is in accordance with some constraints, as Definition 2 states.

The concept of covariant redefinition depends strictly on the *ESI* relation it helps to define, so the following two definitions are mutually recursive. In the following definition the notation "C' is subclass of C" ($C' \leq_{ESI} C$) is used to mean that either $C' = C$ or C' is defined by inheriting C by *ESI*.

Definition 2 (Covariant redefinition). Given a class C', subclass of C:

1. the redefinition v: C'_v in C' of the variable v: C_v defined in C is a covariant redefinition iff C'_v is subclass of C_v;
2. the redefinition $m(p$: $P')$: R' in C' of the method $m(p$: $P)$: R defined in C is a covariant redefinition iff P' is subclass of P and R' is subclass of R.

Consider the following notation. Given a class C, $LVS(C)$ is the set of instance variables explicitly defined in C; $VS(C)$ is the set of instance variables which C either defines or inherits; $LMS(C)$ is the set of methods explicitly defined in C; $MS(C)$ is the set of methods which C either inherits or defines explicitly; $\oplus$ is an operator which takes two sets of typed instance variables and returns their *exclusive* union, such that if there is a variable with the same name in both sets, the one of smaller type is taken.

In Definition 3, we call $\mathcal{I}$ the set of indexes $\{1 \ldots n\}$.

Definition 3 (Enhanced strict inheritance, $\leq_{ESI}$). Given a class C' and the set of its direct superclasses $\{C_1, \ldots, C_n\}$, with $n \geq 1$, we say that the *ESI* relation is fulfilled and write $C' \leq_{ESI} \{C_i\}_{i \in \mathcal{I}}$ iff:

1. for each variable name $v \in \{C', C_1, \ldots, C_n\}$
   a. if $v \in LVS(C')$ and $v \in VS(C_j)$ for $j \in \mathcal{J} \subseteq \mathcal{I}$ then v must be a covariant redefinition in C' w.r.t. C_j for each $j \in \mathcal{J}$;
   b. if $v \in VS(C_j)$ for $j \in \mathcal{J} \subseteq \mathcal{I}$ and the set $\mathcal{J}$ contains more than one element and $v \notin LVS(C')$ then a variable $v \in VS(C_k)$ for some $k \in \mathcal{J}$ must exist such that its type is subtype of the type of each variable $v \in VS(C_j)$ for each $j \in \mathcal{J}$;
2. for each method $m \in LMS(C')$ if $m \in MS(C_j)$ for $j \in \mathcal{J} \subseteq \mathcal{I}$, then m must be a covariant redefinition w.r.t. every method m in $MS(C_j)$ for each $j \in \mathcal{J}$.

In Definition 3, notice that, if one of the conditions in case 1 does hold, we have $VS(C') = (\oplus_{\mathcal{I}} VS(C_i)) \oplus LVS(C')$.

Moreover, in case 2, we have $MS(C') = (\cup_{\mathcal{I}} MS(C_i)) \cup LMS(C')$.

Applying *ESI* definition to classes of Fig. 1, we see that

$$\texttt{DIAG\_MAT\_INT} \leq_{ESI} \{\texttt{SQ\_MAT\_INT}, \texttt{TRI\_MAT\_INT}\} \leq_{ESI} \texttt{MAT\_INT} .$$

This shows how naturally the subtyping inheritance mechanism based on a covariant redefinition rule models the idea of class specialization.

In order to evaluate the correspondence between notions of *type/subtyping* and *class/ESI* we note that, given a class C, sets $VS(C)$, $LMS(C)$ remain defined, and we can build the corresponding type T_C as

TYPE T_C:
VARIABLES $VS(C)$
FUNCTIONS {signatures of methods in $LMS(C)$}

which is unusual, w.r.t. record types setting, yet correct if we look at the type model defined in Di Blasio and Temperini (1997).

It is simple to note that if two given classes C_1 and C_2 are in *ESI* relation, then corresponding types are in subtyping relation:

$$C_1 \leq_{ESI} C_2 \Rightarrow T_{C_1} \leq T_{C_2} .$$

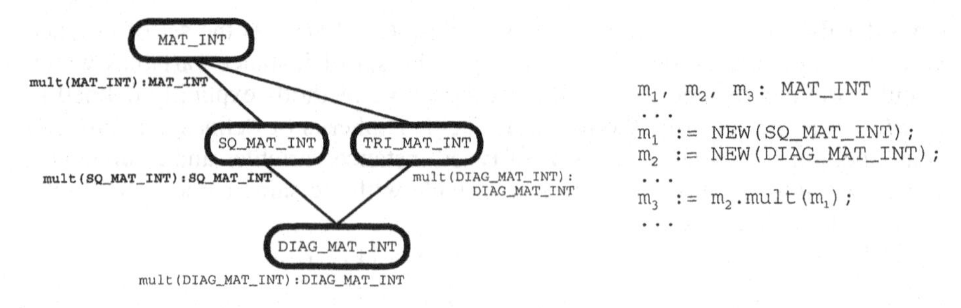

a

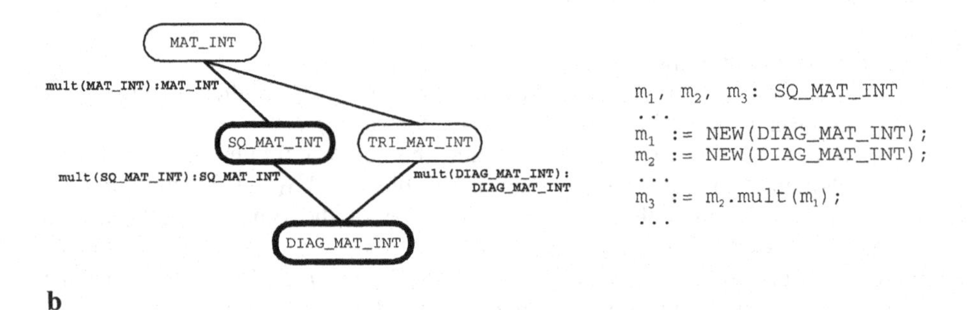

b

Fig. 1. Execution of correct method invocations

Above we have stated some main characteristics of an object-oriented programming language based on *ESI*. In order to make such definition complete, we must specify how correctness of language expressions can be established. This is done by application of the type-checking rules given in Di Blasio and Temperini (1997) and by a suitable method lookup algorithm.

The next section will step into details of the method lookup algorithm. Here we give just a procedural description of such an algorithm, to describe main aspects through an example.

In Fig. 1, the `mult` method defined in `MAT_INT` makes the method invocation of case (a) statically correct; independent of any legal polymorphic assignment of variables m_1 and m_2, it is one of the methods applicable at run-time. Hence, to execute the invocation m_2`.mult(`m_1`)`, we have two applicable methods: those in classes `MAT_INT` and `SQ_MAT_INT`. At run-time, once the method in `DIAG_MAT_INT` has proved unapplicable, if the method lookup algorithm does not stop at `DIAG_MAT_INT`, the most specialized among such applicable methods is executed (that defined in `SQ_MAT_INT`, in this case). In case (b) the `mult` method defined in `SQ_MAT_INT` makes the method invocation statically correct, inferring type `SQ_MAT_INT` for it. At run-time the method lookup algorithm looks for the executable method only in classes `SQ_MAT_INT` and `DIAG_MAT_INT`. Hence method `mult` of class `TRI_MAT_INT`, which is applicable in fact, but out of the searching space, is not considered. It is out of searching space since it is not related with class SQ_MAT_INT on whose subtypes variables m_1, m_2 were established to range.

3 Method lookup algorithm

The previous section gave a short description of the mechanism we propose for retrieving, at run-time, the code to be executed at occurrence of a method invocation.

As the class structure has been stated, at most one implementation of a given method m is defined in each class. So the method lookup can be interpreted as a searching procedure for applicable methods, within the class hierarchy. It is simple to note that in Definition 1, a class may inherit more than one method, with the same name, from different superclasses. For each method invocation, depending on the actual parameter, only one of them will be chosen by the method lookup algorithm. This situation does not occur while inheriting variables. This is the reason why a variable inherited by several direct superclasses, with incompatible definitions, must be redefined properly.

Note that in case of *Single ESI* always a unique most specialized executable method exists, once the method invocation was correct. This is because, by covariant redefinition rule, always a bottom element exists in the set of applicable methods. However, this is not the case of *Multiple ESI*. While dealing with multiple inheritance, several uncomparable applicable methods might exist. This is the case when the set of applicable methods has no unique bottom element.

This means that, given a syntactically correct method invocation o.m(q) over a variable o declared of class C, with argument q:Q, we can retrieve either a unique most specialized method ($\bar{\mathtt{m}}$) or a set of equivalently selectable ones, taken out from a suitable set of classes ($\overline{M}$).

The method $\bar{\mathtt{m}}$ or the set $\overline{M}$ are detected in the set of classes which are superclasses of C′ (the actual class of object o, where $\mathtt{C}' \leq_{ESI} \mathtt{C}$) and subclasses of $\mathtt{C}_0$ (the superclass of C which makes o.m(q) type correct).

However, in order to answer the method invocation, only one executable method must be selected. If a unique $\bar{\mathtt{m}}$ is given as above, then it is the selected method; otherwise a further ordering criterion, say K, is needed in order to state from which class, in the set $\overline{M}$, the selected method must be taken.

The algorithm Method Lookup, given a type correct method invocation o.m(q), performs a breadth first search in a subgraph of the class hierarchy, and returns the most specific executable method (w.r.t. K) defined in a superclass of the class of o. This is done by choosing one out of the classes in set $\overline{M}$, constructed by the algorithm.

Algorithm Method Lookup
Input: a correct method invocation $o.m(q)$ with o: ACTUAL, q: Q;
the class C_0, which made $o.m(q)$ statically correct;
a criterion K;
Output: the most specialized executable method $\bar{m}$.
  QUEUE := ∅;

```
  M̄ := ∅;
  enqueue(ACTUAL, QUEUE);
  while not empty-queue(QUEUE) do
    begin
      C := front(QUEUE);
      dequeue(QUEUE);
      if (m(p: P): R is defined in C, s.t. Q ≤_ESI P)
        then if (∀C' ∈ M̄, C' ≰_ESI C)
            then cons(C, M̄);
          endif;
        else for all C' ∈ DS(C, C_0) do
            if (C' ∉ QUEUE) and (∀C'' ∈ QUEUE, C'' ≰_ESI C')
            then enqueue(C', QUEUE);
            endif;
      endif;
    end;
  return m̄ := K(M̄);
end algorithm.
```

We note that the presented algorithm is redundant, since it provides the whole set of minimal classes in which applicable methods are defined. This set $\overline{M}$ is detectable within the subgraph bounded by the class which allowed to declare the method invocation correct (upper) and the class of the actual message receiver (lower). However, we applied a criterion for our implementation, such that the first coming into $\overline{M}$ is the class from which the executable method is taken.

In algorithm Method Lookup QUEUE is a queue of classes, to be inspected while looking for applicable methods within the searching space of Method Lookup. The notation $DS(C, C_0)$ stands for the set of direct superclasses of the class C, which in turn are subclasses of C_0.

Although an *ordering by level* is preserved, the ordering among classes of the same level is the *inheritance order*, that is the one derived by *DS* set.

The following example shows the behavior of the algorithm in a more general case of multiple inheritance. Suppose we have the hierarchy of Fig. 2, and want to perform the method invocation `o.m(q)`, where `o`: $\mathtt{C}_0$ refers to an object of class ACTUAL, and `q` is declared of class `Q`. Moreover, suppose that $\mathtt{Q} \leq_{ESI} \{\mathtt{P}_3, \mathtt{P}_5, \mathtt{P}_7\}$, and $\mathtt{Q} \not\leq_{ESI} \{\mathtt{P}_2, \mathtt{P}_4, \mathtt{P}_\mathtt{A}\}$, in order to evaluate the applicability of methods.

First we have QUEUE = {ACTUAL} and $\overline{M} = \emptyset$. During the first iteration of the while loop, ACTUAL is assigned to C and dequeued; no method m is applicable in ACTUAL, so its direct superclasses are enqueued: QUEUE = $\{\mathtt{C}_4, \mathtt{C}_7, \mathtt{C}_6\}$, $\overline{M} = \emptyset$. Note that $\mathtt{C}_8$ is not enqueued, since it is not a subclass of C_0.

At the second iteration, $C = \mathtt{C}_4$ and the method `m` in C is not executable: $\mathtt{C}_2, \mathtt{C}_3$ are enqueued.

Then $\mathtt{C}_7$ is assigned to C and dequeued. Its method `m` is applicable, so $\overline{M}$ becomes $\{\mathtt{C}_7\}$.

After the third iteration we have QUEUE = $\{\mathtt{C}_6, \mathtt{C}_2, \mathtt{C}_3\}$ and $\overline{M} = \{\mathtt{C}_7\}$.

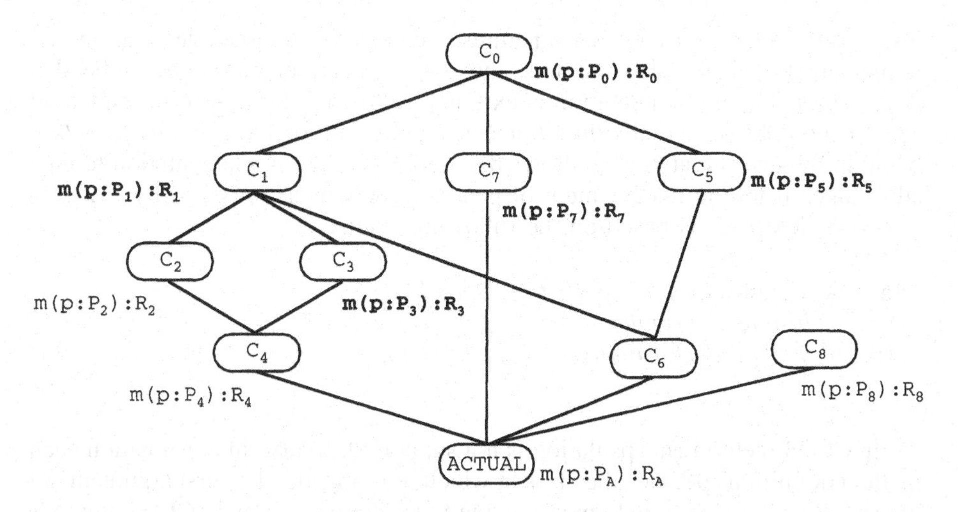

Fig. 2. Method Lookup on multiple inheritance hierarchy: applicable methods in boldface

Then $C = \mathtt{C}_6$ has no method `m` at all, and $\mathtt{C}_1, \mathtt{C}_5$ are enqueued: QUEUE $= \{\mathtt{C}_2, \mathtt{C}_3, \mathtt{C}_1, \mathtt{C}_5\}$.

During the fifth iteration, $C = \mathtt{C}_2$ and the method `m` in C is not executable: $\mathtt{C}_1$ is not enqueued since it is already in QUEUE.

At the sixth iteration an applicable method is detected in $\mathtt{C}_3$, which is added to $\overline{M}$: QUEUE $= \{\mathtt{C}_5, \mathtt{C}_1\}$ and $\overline{M} = \{\mathtt{C}_7, \mathtt{C}_3\}$.

At the seventh iteration $\mathtt{C}_5$ is added to $\overline{M}$. Then $\mathtt{C}_1$ is inspected, checking one more applicable method, but $\overline{M}$ is not increased, since it already contains $\mathtt{C}_3 \leq_{ESI} \mathtt{C}_1$. Now QUEUE is empty, and the procedure stops returning the method selected by the application of K to $\overline{M} = \{\mathtt{C}_7, \mathtt{C}_3, \mathtt{C}_5\}$.

4 Implementation of *ESI* in TASSO-L

The *ESI* mechanism has been implemented within a prototype programming system. This implementation has been developed in order to experiment with different solutions for the static and dynamic treatment of *ESI*-based class hierarchies. So we decided to implement only the most significant aspects, installing them in CLOS. By exploiting the Metaobject Protocol of CLOS, we have implemented a wide set of modifications to the basic language features. Many independent language characteristics are provided: support of *ESI*, with related syntactic checks at class-definition time; use of a different class structure, based on abstract data types approach, with encapsulation of method declarations; dynamic treatment of *ESI*-based programs, through the implementation of the method lookup algorithm. Moreover a static type checking is provided.

In the following we give the main syntactic characteristics of the present implementation of TASSO-L.

The provided class structure corresponds to Definition 1. It encapsulates instance variables and method signatures. Actually we separate declaration and implementation parts of a method, and only the former has to be embedded in the class. The implementation is external, following the typical multimethod approach of CLOS. Actually this choice is not relevant for static treatment of *ESI*, while in dynamic treatment it allows the method lookup implementation to take advantages from the management of generic functions as provided by CLOS.

Class definition is based on the following syntax:

(define-class⟨*class-name*⟩ ⟨*list-of-superclasses*⟩
 ⟨*list-of-instance-variables*⟩
 ⟨*list-of-method-declarations*⟩
)

In CLOS method encapsulation is not supported, so also in our notation each method definition specifies the class in which it is embedded as first argument. At present this is needed to help mapping methods in our notation to CLOS methods naturally. However, such constraint will be eliminated in further versions.

So declaring a method $\texttt{m}(p_1\texttt{:}\ T_1\texttt{;}\ p_2\texttt{:}\ T_2\texttt{;}\ \ldots\texttt{;}\ P_n\texttt{:}\ T_n)\texttt{:R}$ within class T_1 means to encapsulate it (and its implementation) into T_1. Hence the functional form $\texttt{m}(o_1,o_2,\ldots,o_n)$ stands for method invocation, corresponding to the dot-notation $o_1.\texttt{m}(o_2,\ldots,o_n)$ that we have used in the previous sections.

An example shows what ⟨*class-name*⟩, ⟨*list-of-superclasses*⟩, ⟨*list-of-instance-variables*⟩, ⟨*list-of-method-declarations*⟩ represent:

```
(define-class ring ()
()
((sum ('ring 'ring) 'ring)
(prod ('ring 'ring) 'ring))
)

(define-class rationale (ring)
((numerator 'integer-type)
(denominator 'integer-type))
((sum ('rationale 'rationale) 'rationale)
(prod ('rationale 'rationale) 'rationale)
(read-obj ('rationale) 'rationale)
(print-obj ('rationale) 'rationale))
)
```

In this example, the class `rationale`, that models rational numbers, inherits all attributes from the class `ring`, specializing some of them by covariant redefinition and adding other ones. Note that the class `integer-type` is a primitive class in our system.

At class-definition time we check covariant redefinition constraints on instance variables and method declarations, according to *ESI*.

As previously specified, the complete method definition is external to the class, and it is based on the syntax below, following method definition of CLOS.

`(define-method` ⟨*method-name*⟩ ⟨*list-of-specialized-parameters*⟩ ⟨*body*⟩`)`

⟨*list-of-specialized-parameters*⟩ ::= `((`⟨*first-param-name*⟩
⟨*first-param-type*⟩`)` {`(`⟨*param-name*⟩ ⟨*param-type*⟩`)`}`)`;

⟨*body*⟩ ::= [⟨*declarative-form*⟩] {⟨*instruction*⟩};

For instance, let us consider the following code, implementing the addition method for rational numbers:

```
(define-method sum ((a rationale) (b rationale))
  declare-local-var (
    (c rationale)
    (n1 integer-type)
    (n2 integer-type)
    (d1 integer-type)
    (d2 integer-type)
    )
  setq n1 (access-instance 'numerator a))
  setq d1 (access-instance 'denumerator a))
  setq n2 (access-instance 'numerator b))
  setq d2 (access-instance 'denumerator b))
  setq c (make-object 'rationale
    '(numerator (sum (prod n1 d2) (prod n2 d1)))
    '(denumerator (prod d1 d2))
  ) c))
```

About the declarative forms, we stress that type declaration is mandatory for local variables and method formal parameters.

The "instruction" part of the body is composed basically by the following kinds of statements: ⟨*assignment-expr*⟩, ⟨*make-expr*⟩, ⟨*execute-expr*⟩, ⟨*read-value-expr*⟩, ⟨*loop-expr*⟩, ⟨*if-expr*⟩.

While ⟨*loop-expr*⟩ and ⟨*if-expr*⟩ adhere to the usual Lisp syntax, some details about the other expressions are provided in the following.

A ⟨*make-expr*⟩ is responsible for the instantiation of an object of given class, through the execution of a statement `make-object`. Optionally, initialization of instance variables is allowed.

Object instantiation is achieved as follows:

⟨*make-expr*⟩ ::= `(make-object` ⟨*quoted-class-name*⟩
{`(`⟨*inst-var-name*⟩ ⟨*obj-exp*⟩`)`}`)`;

⟨*obj-exp*⟩ ::= ⟨*make-expr*⟩ | ⟨*execute-expr*⟩ | ⟨*variable-name*⟩ |
| ⟨*read-value-expr*⟩;

The prototype supports reading and writing access to instance variables of objects. Note that writing is a clear violation of the encapsulation principle, so we limit its usage.

Access expressions are as follows:

⟨read-value-expr⟩ ::=
  `(read-value-from-instance-variable`
  ⟨quoted-inst-var-name⟩ *⟨read-obj-exp⟩*`);`

⟨read-obj-exp⟩ ::= *⟨make-expr⟩* | *⟨execute-expr⟩* | *⟨variable-name⟩* |
  | *⟨parameter-name⟩* | *⟨read-value-expr⟩*;

⟨assign-slot-expr⟩ ::=
  `(assign-instance-variable`
  ⟨inst-var-name⟩ *⟨variable-name⟩* *⟨obj-exp⟩*`);`

The meaning of the last two expressions, in terms of dot-notation, is the following:

`(read-value-from-instance-variable attr o)` corresponds to `o.attr`;
`(assign-instance-variable v o x)` corresponds to `x := o.v`.

Any access to instance variables of an object is allowed, via the previous statements, only within the body of the encapsulated methods. This checking is performed statically.

The usual assignment expression is as follows:

⟨assignment-expr⟩ ::= *⟨setq-expr⟩* | *⟨assign-slot-expr⟩*;
⟨setq-expr⟩ ::= `(setq` *⟨variable-name⟩* *⟨obj-exp⟩*`);`

To perform static type checking of language instructions, we have implemented a static type inference mechanism following rules of Di Blasio and Temperini (1997). In particular, checks are performed on each assignment statement, to verify that the declared class for the variable or the instance variable is superclass of the one statically inferred for the right-value of the assignment. In each method invocation the static correctness is verified, and is inferred the class which is the upper bound of the Method Lookup searching space.

Method invocation is achieved through a statement `execute-method`, whose syntax is as follows:

⟨execute-expr⟩ ::=
  `(execute-method` *⟨method-name⟩* *⟨list-of-actual-param⟩*`);`

⟨list-of-actual-param⟩ ::= *⟨read-obj-exp⟩*{*⟨obj-exp⟩*};

In the present implementation, the method lookup mechanism is as designed in Sect. 3. Given a compile-time correct method invocation we infer the class which represents the upper bound of the run-time searching space. This information is passed to the run-time mechanism, making a substitution as follows:

$$\texttt{(execute-method m } p_1\ p_2\ \ldots\ p_n\texttt{)}$$
$$\longrightarrow$$
$$\texttt{(exec-method m } T_1\ p_1\ p_2\ \ldots\ p_n\texttt{)}$$

T_1 being the statically inferred class, which represents the upper vertex of the run-time searching space for that method invocation.

The implemented method lookup algorithm starts from the construction of a list of applicable methods, in most-specific-first order, calling the suitable internal function of CLOS. This list contains a method if and only if such method is defined in a superclass of the lower vertex of the diamond. Such lower vertex is the actual class of the first parameter of the (functional-style) method invocation. So, looking for the most specialized method is easily simulated by scanning the list of applicable methods from the least to the most specialized, searching for the first method whose first parameter has a declared type which is subtype of the upper vertex class. We remark that the method precedence order of CLOS has been used as criterion to determine the most specialized method.

5 Experiments with TASSO-L

In order to run TASSO-L programs, the set of new definitions needed by our modifications must be loaded by CLOS system. Then the TASSO-L program can be loaded and its execution is launched by call of a special function `main`, that is attached to all TASSO-L programs. Before execution the program can be statically type-checked, in a preprocessing phase (formerly we called it `pre-compilation` and this is still the name used in the examples). During this preprocessing the TASSO-L code is type-checked and the search space for Method Lookup of each method invocation is set up, following the rules in Sect. 3. In the following, we recall some of the examples in previous sections, in order to experiment with the implementation.

First we define the necessary set of classes as follows. Notice that definitions are incomplete, since they are used here just as examples.

```
(define-class ring ()
()
((mult ('ring 'ring) 'ring))
())

(define-class mat-gen ()
((internal-mat-repr 'array-type)
(numrow 'integer-type)
(numcol 'integer-type))
()
())

(define-class square-mat (mat-gen)  ()  ()  ())

(define-class tri-mat-ring (ring mat-gen)  ()  ()  ())

(define-class mat-int (ring mat-gen)
()
((read-obj ('mat-int) 'mat-int)
(print-obj ('mat-int) 'mat-int)
```

```
(prod ('mat-int 'mat-int) 'mat-int))
())
(define-class sq-mat-int (mat-int square-mat)
()
((mult ('sq-mat-int 'sq-mat-int) 'sq-mat-int))
())

(define-class tri-mat-int (mat-int tri-mat-ring)
()
((mult ('tri-mat-int 'tri-mat-int) 'tri-mat-int))
())
```

Class `diag-mat-int` will be defined in Sect. 5.1, in order to show the management of *ESI* relations by our system. Section 5.2 will show the behavior of the system during the phase of static analysis. Section 5.3 will show a variety of error messages produced through the static type checking on some (purposely wrong) example.

5.1 Errors at class-definition time

Suppose the following (wrong) definition for class `diag-mat-int`:

```
(define-math-class diag-mat-int (tri-mat-int sq-mat-int)
()
((mult ('diag-mat-int 'diag-mat-int) 'sq-mat-int)) ())
```

Once all the definitions are in file `es1.tas` we call an operation of *precompilation* to check whether *ESI* rules are respected or not:

```
> (precompilation "es1.tas")
```

here is the output in this case:

```
    ESI VERIFICATIONS IN PROGRESS....
;;; Loading es1.tas
Error: ESI not respected declaring method PROD in class DIAG-MAT-
INT.
...
```

So errors at class-definition time are detected during this precompilation, while the program file is loaded into CLOS system. At present precompilation stops once an error is detected: so only the first occurring error is reported.

5.2 Static analysis

After precompilation a phase of *static analysis* follows. This phase performs the static type checking of *ESI* and provides the upper vertices for the involved method invocations. Once this second phase is successful, the program can be correctly executed by the modified run-time system of CLOS.

In the following, an example is given, implementing the method invocation of Fig. 1 b.

The following class is added to previous definitions:

```
(define-class diag-mat-int (tri-mat-int sq-mat-int)  ()  () ())
```

The code of example must be implemented in a `main` function.

```
define-main ()
(declare-variables (m1 m2 m3)
(declare-variable-type
(m1 'sq-mat-int) (m2 'sq-mat-int) (m3 'sq-mat-int))
(setq m1 (make-math-object 'diag-mat-int))
(setq m2 (make-math-object 'diag-mat-int))
(setq m3 (execute-method 'mult m1 m2))))
(setq x (make-math-object 'main))
```

Once the precompilation has been successfully executed, the static analysis phase follows:

```
> (precompilation "es2.tas" 'minimal)
ESI VERIFICATIONS IN PROGRESS....
;;; Loading es2.tas
;;; Finished loading es2.tas
STATIC ANALYSIS IN PROGRESS

T
>
```

In this case the program is correct; it can be executed by calling the function `main`:

```
>  (main)
```

Output is:

```
WARNING: execution of the virtual method PROD in the class SQ-MAT-
INT.
#i<a SQ-MAT-INT 7300432 >
>
```

since actually the method `mult` defined in `sq-mat-int` is only declared and not implemented.

Note that even if we reverse the order of inherited classes in the definition of `diag-mat-int`,

```
(define-class diag-mat-int (sq-mat-int tri-mat-int)  ()  ()  ())
```

the output is exactly the same, since the method lookup of CLOS has been completely overridden.

5.3 Errors in static analysis and execution

The following classes are defined for this example:

```
(define-class ring ()
()
((sum ('ring 'ring) 'ring)
(prod ('ring 'ring) 'ring))
)

(define-class rationale (ring)
((numerator 'integer-type)
(denominator 'integer-type))
((sum ('rationale 'rationale) 'rationale)
(prod ('rationale 'rationale) 'rationale)
(correct ('rationale) 'boolean-type)
(zero-number-p ('rationale) 'boolean-type)
(equal-obj ('rationale 'rationale) 'boolean-type)
(copy-obj ('rationale) 'rationale)
(read-obj ('rationale) 'rationale)
(print-obj ('rationale) 'rationale)
(ERROR-METH ('rationale 'rationale) 'rationale))
)
```

Methods are defined and implemented in this case, but here we point out only the implementation of method ERROR-METH, used for collecting a set of interesting errors:

```
(define-method ERROR-MET ((a rationale)(b rationale))
(declare-local-variables (x1 x2 x3)
(declare-type (x1 'rationale)(x2 'rationale)(x3 'ring))
(setq non_decl_var (make-math-object 'rationale))
(loop
(setq x1 (make-math-object 'rationale))
(if x1
(assign-instance-variable 'notaslot x1 (execute-method 'sum x3 x3))
(read-value-from-instance-variable 'notaslot x1))
(setq x2
(execute-method 'sum
(execute-method 'sum x1 x2)
(make-math-object 'ring))) )
(assign-instance-variable 'numerator non_decl_var x1)
(assign-instance-variable 'numerator x1
(execute-method 'prod
```

```
(read-value-from-instance-variable 'numerator x2)
(make-math-object 'real-type)))
))
```

The output of precompilation is given in Fig. 3. No errors are detected during the verification of inheritance relation. Here we point out the most significant errors detected during static analysis.

Instruction

```
(setq x2
(execute-method 'sum
(execute-method 'sum x1 x2)
(make-math-object 'ring)))
```

in loop, is wrong for

```
***ERROR: in the assignment (SETQ X2
(EXEC-METHOD 'SUM 'RING
(EXEC-METHOD 'SUM 'RATIONALE X1
X2)
(MAKE-MATH-OBJECT 'RING))) in method ERROR-MET in class RATIONALE,
the assigned value is not subtype of the variable to be assigned.
```

because the static analysis infers rational type for the result of adding rationals `x1`, `x2`, and ring type for the addition of a rational to a ring; since `ring` is not a subtype of rational, an error is raised.

Another error is detected on the method invocation

```
(assign-instance-variable 'numerator x1
(execute-method 'prod
(read-value-from-instance-variable 'numerator x2)
(make-math-object 'real-type)))
```

with signal

```
***ERROR: bad method invocation: (EXECUTE-METHOD 'PROD
(READ-VALUE-FROM-INSTANCE-VARIABLE
'NUMERATOR X2)
(MAKE-MATH-OBJECT 'REAL-TYPE))
```

This is because no applicable method does exist in this case (in our system no relation is established between the basic types `integer-type` and `real-type`).

6 Conclusions

In this paper we presented a subtyping inheritance mechanism (*ESI*) for an object-oriented programming language, suitable for the specification of data structures and algorithms of a symbolic computation system. *ESI* reflects the

```
> (precompilation "es3.tas" 'medium)
ESI VERIFICATIONS IN PROGRESS....
;;; Loading es3.tas
;;; Finished loading es3.tas
STATIC ANALYSIS IN PROGRESS
***ERROR: The variable NON_DECL_VAR in the instruction (SETQ
NON_DECL_VAR
(MAKE-MATH-OBJECT
'RATIONALE)) in method ERROR-MET in class RATIONALE was not de-
clared.
***ERROR: in the assignment (SETQ NON_DECL_VAR
(MAKE-MATH-OBJECT 'RATIONALE)) in method ERROR-MET in class RATIO-
NALE, the assigned value is not subtype of the variable to be as-
signed.
***ERROR: in the assignment (ASSIGN-INSTANCE-VARIABLE 'NOTASLOT X1
(EXECUTE-METHOD 'SUM X3 X3)) in method ERROR-MET in class RATIONALE,
the slot NOTASLOT does not exist in class RATIONALE and in its su-
perclasses.
***ERROR: in the assignment (ASSIGN-INSTANCE-VARIABLE 'NOTASLOT X1
(EXEC-METHOD 'SUM 'RING X3 X3)) in method ERROR-MET in class RATIO-
NALE, the assigned value is not subtype of the slot to be assigned.
***ERROR: in the statement (READ-VALUE-FROM-INSTANCE-VARIABLE
'NOTASLOT X1), the slot NOTASLOT is not a slot of the object
X1.
***ERROR: in the assignment (SETQ X2
(EXEC-METHOD 'SUM 'RING
(EXEC-METHOD 'SUM 'RATIONALE X1
X2)
(MAKE-MATH-OBJECT 'RING))) in method ERROR-MET in class RATIONALE,
the assigned value is not subtype of the variable to be assigned.
***ERROR: the variable NON_DECL_VAR in the instruction
(ASSIGN-INSTANCE-VARIABLE
'NUMERATOR
NON_DECL_VAR
X1) in method ERROR-MET in class RATIONALE was not declared.
***ERROR: in the assignment (ASSIGN-INSTANCE-VARIABLE 'NUMERATOR
NON_DECL_VAR X1) in method ERROR-MET in class RATIONALE, the as-
signed value is not subtype of the slot to be assigned.

***ERROR: bad method invocation: (EXECUTE-METHOD 'PROD
(READ-VALUE-FROM-INSTANCE-VARIABLE
'NUMERATOR X2)
(MAKE-MATH-OBJECT 'REAL-TYPE))
NIL
>
```

Fig. 3. Static type checking

classical view of object-oriented programming: classes are abstract data types implementations; they embed data structures and related methods; message passing is used to invoke methods. The inheritance is based on a covariant method redefinition rule. Through *ESI* a specialization relation among classes is modeled and management of polymorphic instructions is made safe.

The correspondence between the static and dynamic correctness of programs developed with *ESI* is guaranteed by a set of rules for static type checking and by a suitable enhancement of the usual dynamic binding features, namely the method lookup algorithm. After an invocation, the algorithm selects the method to be executed based on the whole set of actual parameters – not only on the object receiving the method invocation. This is clearly a heritage coming from multimethod languages like CLOS: however, we don't simply inherit it in our programming language; we embedded (with modifications) such characteristic, in order to model by inheritance the covariant subtyping relation described by Di Blasio and Temperini (1997). This allows *ESI* to be a mechanism for developing specialization relationship between inherited and inheriting class.

The prototypal implementation of TASSO-L was also described. It has been developed within the frame of the TASSO project and is built to embed the static and dynamic treatment of *ESI*.

TASSO-L is also going to include other features in addition to those reported in Sect. 4. Genericity will be supported, allowing for the definition of parametric classes and for their static actualization. Moreover we are experimenting with dynamic actualization of parametric classes, that is instantiation of objects whose ground type wasn't completely specified at compile-time: in fact, the whole ground type is established at run-time through suitable constructs (Limongelli and Temperini 1992).

Acknowledgement

This work has been partially supported by MURST and by CNR under the project "Sistemi Informatici e Calcolo Parallelo", grant no. 92.01604.69.

References

Bobrow, D. G., DeMichiel, L. G., Gabriel, P., Keene, S. E., Kiczales, G., Moon, D. A. (1988): Common lisp object system specification, X3JI3 doc. 88-002R, ANSI Common Lisp Standard Committee (also in SIGPLAN Not. 23/9).

Di Blasio, P., Temperini, M. (1997): On subtyping in languages for symbolic computation systems. In: Miola, A., Temperini, M. (eds.): Advances in the design of symbolic computation systems. Springer, Wien New York, pp. 164–178 (this volume).

Kiczales, G., des Rivières, J., Bobrow, D. G. (1991): The art of the metaobject protocol. MIT Press, Cambridge, MA.

Limongelli, C., Temperini, M. (1992): Abstract specification of structures and methods in symbolic mathematical computation. Theor. Comput. Sci. 104: 89–107.

Miola, A. (1997): An overview of the TASSO project. In: Miola, A., Temperini, M. (eds.): Advances in the design of symbolic computation systems. Springer, Wien New York, pp. 21–29 (this volume).

Reasoning capabilities

Deduction and abduction using a sequent calculus

G. Cioni, A. Colagrossi, and A. Miola

1 Introduction

A sequent calculus for automated reasoning is a particular sequent calculus that constitutes a single uniform method to perform different types of logical inferences in first order theories.

The inference problems considered in this paper can be formulated as:

- solution of validity problems for well-formed formulas (wff);
- generation of a wff β such that $\alpha \models \beta$, for a given wff α;
- abduction of a wff α such that $\alpha \models \beta$, for a given wff β.

These three inference problems are known in the literature as *verificative*, *generative*, and *abductive* logic problems, respectively.

Verificative problems are generally solved by a refutation mechanism, while generative and abductive problems have been treated by different automated deduction methods based on the inference schemas of Hilbert and of Gentzen. A Hilbert schema is characterized by many axioms and few inference rules (e.g., modus ponens), whereas a Gentzen schema is characterized by few axioms and many rules. The Gentzen schema leads to a natural calculus, being very close to the human way of reasoning (Hermes 1973).

Several methods have been developed in the context of the Hilbert schema, e.g., the resolution method, which is the basis for Prolog (Kowalski 1979). On the other hand, methods based on the Gentzen schema have been studied by Bibel (1987), who proposed a uniformly better method than resolution, and are still under investigation by several researchers to improve the original idea of sequent calculus (Gallier 1986, Girard et al. 1989).

A sequent is a particular logical expression used to model inference problems. A sequent calculus is a proof system which employs a set of predefined rules and a set of properties, to decompose a given sequent into elementary components and to verify the validity of the sequent, by operating on its components. It is well known that this decomposition process does not always terminate, due to the semi-decidability of first-order logic.

Several sequent calculi have been defined and most of them operate by widely using the assumption rule (Ebbinghaus et al. 1984, Hermes 1973), which limits the full automation of the method. For this reason, the most effective sequent calculus appears to be the proof system defined by Gallier (1986), proved to be sound and complete and, moreover, fully automatizable.

Unfortunately, every sequent calculus in the literature is limited to treat only validity problems. It can be noted that, if the proving procedure terminates with an invalid sequent, the process of reasoning can proceed further, with the goal of generating and abducing new formulas, which, when added to the initial sequent, make it valid.

The main objective of this paper is to define an extended sequent calculus to deal with generative and abductive logic problems, as well as verificative problems, within a single methodological and computational approach.

The proposed sequent calculus follows Gallier's (1986) style. However, the need for a uniform approach to deal with verificative, generative, and abductive problems requires a new formulation for the decomposition rules for quantified formulas and for the related proving procedure. These are presented in Sects. 2 and 3.

If the process has to be continued, by reasoning on a given non valid sequent in order to generate and abduce new formulas, the method employs a new set of transformation rules to build sets of formulas, which, once combined in an appropriate way, allow one to compose a valid sequent from the initial one. This is presented in Sect. 4.

The proposed sequent calculus is flexible (indeed, different types of inferences can be treated by a single method), natural (input logical formulas are not required to be expressed in any predefined form and the proof is carried out in a way very close to the human reasoning), and fully automatizable.

In the next sections, the basics of the proposed sequent calculus (notations, definitions, decomposition rules and properties) are given. Then, the method for proving and reasoning with the sequent calculus, namely for verifying the validity of a sequent, and for generating or abducing new formulas is presented. In Sect. 5, we hint at possible applications.

2 Basics of the sequent calculus

2.1 Notations

Lower-case greek letters denote wffs; capital greek letters denote sets of wffs; x, y, and z denote variables; h, i, j, k, l, m, and n denote integers; other lower-case letters denote sequents; Y denotes sets of symbols; P denotes predicate symbols; other capital letters denote sets of sequents.

2.2 Definitions

Let Γ be a set $\{\alpha_1, \ldots, \alpha_n\}$ and Δ be a set $\{\sigma_1, \ldots, \sigma_m\}$ of wffs in first-order logic (CP_1), constructed with the usual connectives and quantifiers $\wedge, \vee, \neg, \supset, \forall, \exists$. A *sequent* is a logical expression written as

$$\Gamma \Rightarrow \Delta \quad \text{or} \quad \alpha_1, \ldots, \alpha_n \Rightarrow \sigma_1, \ldots, \sigma_m \tag{1}$$

where Γ is called the *antecedent*, Δ is called the *succedent* and the formulas $\alpha_1, \ldots, \alpha_n, \sigma_1, \ldots, \sigma_m$ are called the *component formulas*. The antecedent and succedent of a sequent may be empty.

A sequent is *empty* if both the antecedent and the succedent are empty.

A sequent is *elementary* if its component formulas are literals.

A sequent is *open* if its component formulas are quantifier-free.

An open sequent is *A-open* if its succedent is empty.

An open sequent is *S-open* if its antecedent is empty.

An *A-clause* is a conjunction of literals; an *S-clause* is a disjunction of literals; a *clause* is either an A-clause or an S-clause.

The antecedent (succedent) of a sequent, as in (1), is *valid*, if $\alpha_1 \wedge \alpha_2 \wedge \ldots \wedge \alpha_n$ $(\sigma_1 \vee \sigma_2 \vee \ldots \vee \sigma_m)$ is valid; it is a *contradiction*, if $\alpha_1 \wedge \alpha_2 \wedge \ldots \wedge \alpha_n$ $(\sigma_1 \vee \sigma_2 \vee \ldots \vee \sigma_m)$ is a contradiction; it is *falsifiable*, otherwise.

A sequent is *valid* if the formula

$$\alpha_1 \wedge \ldots \wedge \alpha_n \supset \sigma_1 \vee \ldots \vee \sigma_m \tag{1'}$$

is valid; a sequent is a *contradiction* if (1′) is a contradiction; a sequent is *falsifiable* if it is neither valid nor a contradiction.

A sequent s_1 is *equivalent* to a sequent s_2 if, for any interpretation, either s_1 and s_2 are both valid or both contradictory.

2.3 Decomposition rules for logical connectives

Let $\mathbf{R}_\mathrm{D}^8 = \{\underline{\wedge\mathbf{A}}, \underline{\wedge\mathbf{S}}, \underline{\neg\mathbf{A}}, \underline{\neg\mathbf{S}}, \underline{\vee\mathbf{A}}, \underline{\vee\mathbf{S}}, \underline{\supset\mathbf{A}}, \underline{\supset\mathbf{S}}\}$ be the set of eight decomposition rules to be applied to the antecedent and the succedent of sequents including formulas whose connectives are $\wedge, \neg, \vee, \supset$, respectively. These rules are defined as follows:

$\underline{\wedge\mathbf{A}}$: from the sequent $\Gamma_1, \alpha' \wedge \alpha'', \Gamma_2 \Rightarrow \Delta$ the sequent

$$\Gamma_1, \alpha', \alpha'', \Gamma_2 \Rightarrow \Delta$$

is derived;

$\underline{\wedge\mathbf{S}}$: from the sequent $\Gamma \Rightarrow \Delta_1, \sigma' \wedge \sigma'', \Delta_2$ the sequents

$$\Gamma \Rightarrow \Delta_1, \sigma', \Delta_2 \;,$$
$$\Gamma \Rightarrow \Delta_1, \sigma'', \Delta_2$$

are derived;

$\underline{\neg\mathbf{A}}$: from the sequent $\Gamma_1, \neg\alpha, \Gamma_2 \Rightarrow \Delta$ the sequent

$$\Gamma_1, \Gamma_2 \Rightarrow \alpha, \Delta$$

is derived;

$\underline{\neg\mathbf{S}}$: from the sequent $\Gamma \Rightarrow \Delta_1, \neg\sigma, \Delta_2$ the sequent

$$\sigma, \Gamma \Rightarrow \Delta_1, \Delta_2$$

is derived;

$\underline{\vee\mathbf{A}}$: from the sequent $\Gamma_1, \alpha' \vee \alpha'', \Gamma_2 \Rightarrow \Delta$ the sequents

$$\Gamma_1, \alpha', \Gamma_2 \Rightarrow \Delta\ ,$$
$$\Gamma_1, \alpha'', \Gamma_2 \Rightarrow \Delta$$

are derived;

$\underline{\vee\mathbf{S}}$: from the sequent $\Gamma \Rightarrow \Delta_1, \sigma' \vee \sigma'', \Delta_2$ the sequent

$$\Gamma \Rightarrow \Delta_1, \sigma', \sigma'', \Delta_2$$

is derived;

$\underline{\supset\mathbf{A}}$: from the sequent $\Gamma_1, \alpha' \supset \alpha'', \Gamma_2 \Rightarrow \Delta$ the sequents

$$\Gamma_1, \Gamma_2 \Rightarrow \alpha', \Delta\ ,$$
$$\Gamma_1, \alpha'', \Gamma_2 \Rightarrow \Delta$$

are derived;

$\underline{\supset\mathbf{S}}$: from the sequent $\Gamma \Rightarrow \Delta_1, \sigma' \supset \sigma'', \Delta_2$ the sequent

$$\sigma', \Gamma \Rightarrow \Delta_1, \sigma'', \Delta_2$$

is derived.

The set of the first four decomposition rules is the minimum set of rules which allows one to operate on any sequent whose formulas are quantifier-free in CP_1. The last four rules have been introduced to avoid unnecessary restrictions on the language.

2.4 Decomposition rules for quantifiers

Let $\mathbf{R}_{\mathrm{D}}^4 = \{\underline{\exists\mathbf{A}}, \underline{\exists\mathbf{S}}, \underline{\forall\mathbf{A}}, \underline{\forall\mathbf{S}}\}$ be the set of four decomposition rules to be applied to the antecedent and the succedent (of a rectified sequent) including quantified formulas.

Let s be a sequent; let $\Theta = \{(y_0, \Phi_0), (y_1, \Phi_1), \ldots, (y_k, \Phi_k)\}$ be a set of pairs, where y_i, for $i = 0, \ldots, k$, are constants or variables occurring free in s and Φ_i, for $i = 0, \ldots, k$, are sets of quantified formulas occurring in s; let $Y = \{y_{k+1}, y_{k+2}, \ldots\}$ be a set of distinct symbols not occurring in s.

In the following decomposition rules the sets Θ and Y are used to correctly instantiate the quantified variables of a given sequent (i.e., to keep track of the various instantiations). The four rules are defined as follows:

<u>∃A</u>: from the sequent $\Gamma_1, \exists x\alpha, \Gamma_2 \Rightarrow \Delta$ the sequent

$$\Gamma_1, \alpha[y_i/x], \Gamma_2 \Rightarrow \Delta$$

is derived, where $y_i \in Y$; the symbol y_i is removed from the set Y and the pair $(y_i, \emptyset)$ is added to the set Θ;

<u>∃S</u>: from the sequent $\Gamma \Rightarrow \Delta_1, \exists x\sigma, \Delta_2$ the sequent

$$\Gamma \Rightarrow \exists x\sigma, \Delta_1, \sigma[y_{l_0}/x], \ldots, \sigma[y_{l_k}/x], \Delta_2$$

is derived, where $y_{l_0}, \ldots, y_{l_k}$ are the first elements of the pairs (y_{l_0}, Φ_{l_0}), $\ldots$, (y_{l_k}, Φ_{l_k}) in Θ such that $\exists x\sigma$ doesn't belong to any of the $\Phi_{l_0}, \ldots,$ Φ_{l_k}; the set Θ is modified by adding $\exists x\sigma$ to $\Phi_{l_0}, \ldots, \Phi_{l_k}$;

<u>∀A</u>: from the sequent $\Gamma_1, \forall x\alpha, \Gamma_2 \Rightarrow \Delta$ the sequent

$$\forall x\alpha, \Gamma_1, \alpha[y_{l_0}/x], \ldots, \alpha[y_{l_k}/x], \Gamma_2 \Rightarrow \Delta$$

is derived, where $y_{l_0}, \ldots, y_{l_k}$ are the first elements of the pairs (y_{l_0}, Φ_{l_0}), $\ldots$, (y_{l_k}, Φ_{l_k}) in Θ such that $\forall x\alpha$ doesn't belong to any of the $\Phi_{l_0}, \ldots,$ Φ_{l_k}; the set Θ is modified by adding $\forall x\alpha$ to $\Phi_{l_0}, \ldots, \Phi_{l_k}$;

<u>∀S</u>: from the sequent $\Gamma \Rightarrow \Delta_1, \forall x\sigma, \Delta_2$ the sequent

$$\Gamma \Rightarrow \Delta_1, \sigma[y_i/x], \Delta_2$$

is derived, where $y_i \in Y$; the symbol y_i is removed from the set Y and the pair $(y_i, \emptyset)$ is added to the set Θ.

$\mathbf{R}_{\mathrm{D}}^{12} = \mathbf{R}_{\mathrm{D}}^{8} \cup \mathbf{R}_{\mathrm{D}}^{4}$ is the complete set of decomposition rules.

A sequent is *atomic* if none of the decomposition rules of $\mathbf{R}_{\mathrm{D}}^{12}$ applies.

2.5 Properties of a sequent

As far as the semantics of a sequent is concerned, several properties hold:

i. A sequent s is equivalent to any sequent obtained by rearranging the order of the component formulas in the antecedent and in the succedent of s.
ii. A sequent s with two equivalent component formulas, either in the antecedent or in the succedent, is equivalent to the sequents obtained by eliminating one of the equivalent formulas from the sequent s.
iii. The sequent $\Gamma_1, \alpha, \Gamma_2 \Rightarrow \Delta$, where α is a tautology, is equivalent to $\Gamma_1, \Gamma_2 \Rightarrow \Delta$.
iv. The sequent $\Gamma \Rightarrow \Delta_1, \sigma, \Delta_2$, where σ is a contradiction, is equivalent to $\Gamma \Rightarrow \Delta_1, \Delta_2$.
v. The sequent $\Gamma_1, \alpha, \Gamma_2 \Rightarrow \Delta$, where α is a contradiction, is valid.
vi. The sequent $\Gamma \Rightarrow \Delta_1, \sigma, \Delta_2$, where σ is a tautology, is valid.
vii. A sequent with one component formula in the antecedent equivalent to one of the component formulas in the succedent, is valid.

A sequent is *basic-valid* if one of the Properties (v), (vi) and (vii) holds.

viii. For any valid sequent $\Gamma \Rightarrow \Delta$ and for any wff σ the sequent $\Gamma \Rightarrow \Delta, \sigma$ is valid.
ix. For any valid sequent $\Gamma \Rightarrow \Delta$ and for any wff α the sequent $\alpha, \Gamma \Rightarrow \Delta$ is valid.
x. Every sequent derived by applying any rule of $\mathbf{R}_{\mathrm{D}}^{12}$ to a valid sequent is valid.
xi. A sequent from which valid sequents are derived by applying any rule of $\mathbf{R}_{D}^{12}$ is valid.
xii. The order of application of the rules of $\mathbf{R}_{\mathrm{D}}^{12}$ to a sequent is immaterial.

Proofs are omitted because they are straightforward consequences of the definitions, sometimes based on a long case analysis.

3 Proving with the sequent calculus

A constructive solution to the validity problem of a given sequent is based on successive applications of the rules of $\mathbf{R}_{\mathrm{D}}^{12}$ to the given sequent to obtain a set of basic-valid sequents. Semi-decision procedures can be defined for the validity problem and the soundness and completeness of the proof system can be stated.

On the basis of Property (xii), several such procedures can be defined, according to different strategies of application of the decomposition rules. The strategy considered here aims at minimizing the number of derived sequents and the number of formulas in the derived sequents. This strategy is defined by associating a priority of application with each of the decomposition rules, as follows, where rules are listed in decreasing order of priority and rules in the same row have the same priority.

$\underline{\exists \mathbf{A}}$, $\underline{\forall \mathbf{S}}$;
$\underline{\forall \mathbf{A}}$, $\underline{\exists \mathbf{S}}$;
$\underline{\wedge \mathbf{A}}$, $\underline{\vee \mathbf{S}}$, $\underline{\supset \mathbf{S}}$;
$\underline{\vee \mathbf{A}}$, $\underline{\wedge \mathbf{S}}$, $\underline{\supset \mathbf{A}}$;
$\underline{\neg \mathbf{A}}$, $\underline{\neg \mathbf{S}}$.

The semi-decision procedure based on this strategy follows. In this procedure the only output parameter of interest for the validity problem of the input sequent s is the set S. Two further output parameters have been introduced, namely the sets Θ and Y'; their use will be clear in Sect. 4.

Procedure $\mathcal{P}$
Input: a rectified sequent s;
Output: a set S of atomic sequents, a set Θ of pairs formed with a symbol and set of quantified formulas, and a set Y' of symbols.

Step 0: *initialize* the sets of sequents K, L and S to the empty set;
if the input sequent has neither constants nor free variables, then *initialize* the set of pairs Θ to the set $\{(y_0, \emptyset)\}$, where y_0 is a symbol not occurring in the input sequent, and the set of symbols Y to the set $\{y_1, y_2, \ldots\}$ of the symbols not occurring in the input sequent, else *initialize* the set of pairs Θ to the set $\{(y_0, \emptyset), (y_1, \emptyset), \ldots, (y_k, \emptyset)\}$, where y_i, for $i = 0, \ldots, k$, are constants or variables occurring free in the input sequent, and the set of symbols Y to the set $\{y_{k+1}, y_{k+2}, \ldots\}$ of the symbols not occurring in the input sequent;
initialize the set of symbols Y' to the set of the first elements of the pairs of the set Θ;

Step 1: *insert* the sequent s into K;

Step 2: repeat
delete all the basic-valid sequents from K;
delete all the atomic sequents from K and *insert* them into S;
delete all the (remaining) sequents from K and *insert* them into L;
repeat
get a sequent r of L including an unmarked formula ρ for which the highest priority rule of $\mathbf{R}_{\mathrm{D}}^{12}$ is applicable, and *apply* the identified rule to the sequent r;
if the applied rule is either $\underline{\forall \mathbf{A}}$ or $\underline{\exists \mathbf{S}}$, then *get* all (unmarked) formulas equal to ρ in the sequents of L other than r, and *apply* the same rule to those sequents;
mark the formula(s) of the sequents obtained by the application of the rule and *insert* the sequent(s) resulting from the previous rule application into L;
until all the formulas of the sequents in L are either marked or atomic;
unmark the formulas of the sequents of L, *insert* the sequents of L into K and *clear* L;
until K is empty.

The following soundness and completeness theorem holds for $\mathbf{R}_{\mathrm{D}}^{12}$ and $\mathcal{P}$.

Theorem 1. The application of $\mathcal{P}$ to a sequent s terminates with an empty output set S iff the sequent s is valid.

The proof is analogous to the one reported by Gallier (1986) so as $\mathbf{R}_{\mathrm{D}}^{12}$ and $\mathcal{P}$ are analogous to the proof system proposed there.

From Theorem 1 and the basics of Sect. 2, the following corollaries can be proved.

Corollary 1. If either the antecedent of a sequent is a contradiction or the succedent of a sequent is valid, then the application of $\mathcal{P}$ to that sequent terminates with an empty set S.

Corollary 2. If α and σ are equivalent wffs, then the application of $\mathcal{P}$ to the sequent $\alpha \Rightarrow \sigma$ terminates with an empty set S.

The following two theorems can also be proved for $\mathcal{P}$.

Theorem 2. Let $S = \{\Gamma_1 \Rightarrow \Delta_1, \ldots, \Gamma_n \Rightarrow \Delta_n\}$ be the non empty output of $\mathcal{P}$ applied to the sequent $\Gamma \Rightarrow \Delta$. Let σ be a wff. The sequent $\Gamma \Rightarrow \Delta, \sigma$ is valid iff all the sequents $\Gamma_i \Rightarrow \Delta_i, \sigma$, for $i = 1, \ldots, n$, are valid.

Proof. Let $\Gamma \Rightarrow \Delta, \sigma$ be a valid sequent. Then, by Property (x), the application of any sequence of rules of $\mathbf{R}_{\mathrm{D}}^{12}$ to the given sequent derives valid sequents. In particular, the sequence of decomposition rules applied to the sequent $\Gamma \Rightarrow \Delta$, during the execution of $\mathcal{P}$, can be also applied to the sequent $\Gamma \Rightarrow \Delta, \sigma$ (i.e., the decomposition rules operate only on Γ or Δ and not on σ) to derive the valid sequents $\Gamma_i \Rightarrow \Delta_i, \sigma$ for $i = 1, \ldots, n$.

Let $\Gamma_i \Rightarrow \Delta_i, \sigma$, for $i = 1, \ldots, n$, be the sequents derived from the sequent $\Gamma \Rightarrow \Delta, \sigma$ by applying the same sequence of decomposition rules applied to the sequent $\Gamma \Rightarrow \Delta$, during the execution of $\mathcal{P}$ (i.e., the decomposition rules operate only on Γ or Δ and not on σ), be valid sequents. Then, by Property (xi), the sequent $\Gamma \Rightarrow \Delta, \sigma$ is valid. □

Theorem 2′. Let $S = \{\Gamma_1 \Rightarrow \Delta_1, \ldots, \Gamma_n \Rightarrow \Delta_n\}$ be the non empty output of $\mathcal{P}$ applied to the sequent $\Gamma \Rightarrow \Delta$. Let α be a wff. The sequent $\alpha, \Gamma \Rightarrow \Delta$ is valid iff all the sequents $\alpha, \Gamma_i \Rightarrow \Delta_i$, for $i = 1, \ldots, n$, are valid.

The proof is analogous to that of Theorem 2.

3.1 An example of proof with the sequent calculus

An example of application of Procedure $\mathcal{P}$ follows. In this example, the formulas obtained by the application of the function *mark* are represented by overlined formulas. The input is the following sequent p:

$$(p) \qquad \forall x_1(P_1(x_1) \supset \neg P_2(x_1)),\ \exists x_2(P_2(x_2) \wedge P_3(x_2)) \Rightarrow$$

The initializations and the execution of the first step give:

$$K = \{p\},\ L = \{\},\ S = \{\},\ \Theta = \{(y_0, \emptyset)\},\ Y = \{y_1, y_2, \ldots\},\ Y' = \{y_0\}\ .$$

At the first iteration, there are neither valid nor atomic sequents, then

$$K = \{\},\ L = \{p\},\ S = \{\}\ .$$

The highest priority rule applicable to p is $\underline{\exists \mathbf{A}}$, then the resulting sequent is:

$$(q) \qquad \forall x_1(P_1(x_1) \supset \neg P_2(x_1)),\ \overline{P_2(y_1) \wedge P_3(y_1)} \Rightarrow$$

and Θ and Y are modified as:

$$\Theta = \{(y_0, \emptyset), (y_1, \emptyset)\},\ Y = \{y_2, y_3, \ldots\}\ .$$

The highest priority rule applicable to unmarked formulas of q is $\underline{\forall \mathbf{A}}$, then the resulting sequent is:

$$(r) \qquad \overline{\forall x_1(P_1(x_1) \supset \neg P_2(x_1))},\ \overline{P_1(y_0) \supset \neg P_2(y_0)},\ \overline{P_1(y_1) \supset \neg P_2(y_1)}\ ,\ \overline{P_2(y_1) \wedge P_3(y_1)} \Rightarrow$$

and Θ is modified as:

$$\Theta = \{(y_0, \{\forall x_1(P_1(x_1) \supset \neg P_2(x_1))\}), (y_1, \{\forall x_1(P_1(x_1) \supset \neg P_2(x_1))\})\}\ .$$

The sequent r is inserted into K and the loop continues.

At this point, $L = \{r\}$ and, even though the highest priority rule is $\underline{\forall \mathbf{A}}$, it cannot be applied. Therefore, the rule $\underline{\wedge \mathbf{A}}$ is applied, producing the sequent:

$$(s) \qquad \forall x_1(P_1(x_1) \supset \neg P_2(x_1)),\ P_1(y_0) \supset \neg P_2(y_0)\ ,\ P_1(y_1) \supset \neg P_2(y_1),\ \overline{P_2(y_1)},\ \overline{P_3(y_1)} \Rightarrow$$

The rule $\underline{\supset \mathbf{A}}$ is applied to s and the results are the following sequents:

$$(t_1) \quad \forall x_1(P_1(x_1) \supset \neg P_2(x_1)),\ \overline{\neg P_2(y_0)},\ P_1(y_1) \supset \neg P_2(y_1),\ \overline{P_2(y_1)},\ \overline{P_3(y_1)} \Rightarrow$$

$$(t_2) \quad \forall x_1(P_1(x_1) \supset \neg P_2(x_1)),\ P_1(y_1) \supset \neg P_2(y_1),\ \overline{P_2(y_1)},\ \overline{P_3(y_1)} \Rightarrow \overline{P_1(y_0)}$$

Then, the same rule $\underline{\supset \mathbf{A}}$ is applied to t_1 and t_2, and the resulting sequents are:

$$(u_1) \quad \forall x_1(P_1(x_1) \supset \neg P_2(x_1)),\ \overline{\neg P_2(y_0)},\ \overline{\neg P_2(y_1)},\ \overline{P_2(y_1)},\ \overline{P_3(y_1)} \Rightarrow$$

$$(u_2) \quad \forall x_1(P_1(x_1) \supset \neg P_2(x_1)),\ \overline{\neg P_2(y_0)},\ \overline{P_2(y_1)},\ \overline{P_3(y_1)} \Rightarrow \overline{P_1(y_1)}$$

$$(u_3) \quad \forall x_1(P_1(x_1) \supset \neg P_2(x_1)),\ \overline{\neg P_2(y_1)},\ \overline{P_2(y_1)},\ \overline{P_3(y_1)} \Rightarrow \overline{P_1(y_0)}$$

$$(u_4) \quad \forall x_1(P_1(x_1) \supset \neg P_2(x_1)),\ \overline{P_2(y_1)},\ \overline{P_3(y_1)} \Rightarrow \overline{P_1(y_0)},\ \overline{P_1(y_1)}$$

The loop continues, with $K = \{u_1, u_2, u_3, u_4\}$. The basic-valid sequents u_1 and u_3 are deleted from K (according to Property (v)), the atomic sequent u_4 is inserted into S and u_2 is inserted into L, then:

$$L = \{u_2\},\ S = \{u_4\}\ .$$

Now, only the rule $\underline{\neg \mathbf{A}}$ is applicable to u_2 and the result is the following atomic sequent v:

$$(v) \qquad \forall x_1(P_1(x_1) \supset \neg P_2(x_1)), P_2(y_1), P_3(y_1) \Rightarrow \overline{P_2(y_0)}, P_1(y_1) .$$

The procedure terminates with:

$$S = \{v, u_4\} ,$$
$$\Theta = \{(y_0, \{\forall x_1(P_1(x_1) \supset \neg P_2(x_1))\}), (y_1, \{\forall x_1(P_1(x_1) \supset \neg P_2(x_1))\})\} ,$$
$$Y' = \{y_0\} ,$$

and the input sequent p is not valid (since S is not empty).

The proof just constructed is shown in its tree form in Fig. 1.

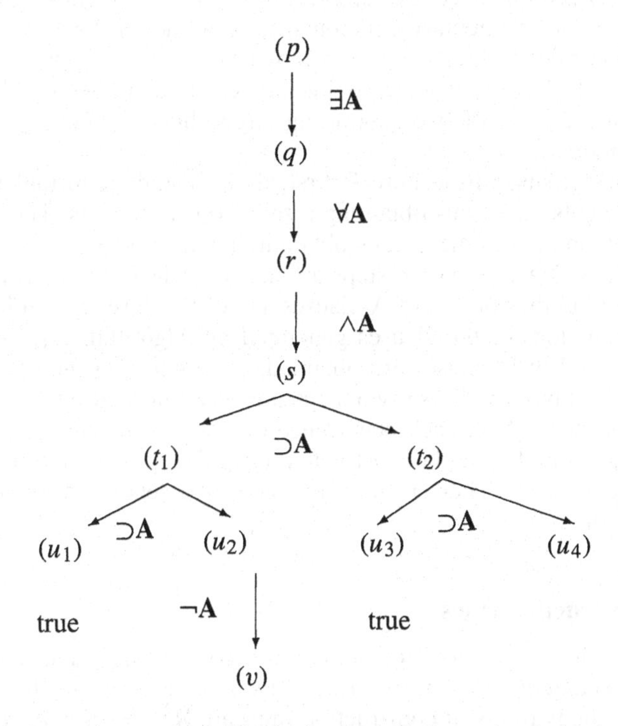

Fig. 1. Proof shown in tree form. Vertical arrows indicate the derivation of sequents by the application of the decomposition rules labelling the arrows; valid sequents are indicated by the constant "true"

4 Reasoning with the sequent calculus

If $\mathcal{P}$ terminates with a non empty output S (i.e., the input sequent is not valid, as in the above example) then, starting from the set S, one can proceed with a further reasoning process on the given sequent. This process consists in constructing a pair of sets $\langle \Sigma^A_{\text{DNF}}, \Sigma^S_{\text{CNF}} \rangle$ of wffs, such that, from the given non valid sequent $\Gamma \Rightarrow \Delta$, for every formula $\sigma \in \Sigma^A_{\text{DNF}}$ and for every formula $\alpha \in \Sigma^S_{\text{CNF}}$, the following valid sequents

$$\Gamma \Rightarrow \Delta, \sigma \tag{2}$$

and

$$\Gamma, \alpha \Rightarrow \Delta \tag{2'}$$

can be generated.

The processes leading to the sequents (2) and (2′) correspond to the inference processes of generation and abduction, respectively.

They are based on successive applications of transformation rules to assemble a sequent from elementary components, as defined in Sect. 4.1. These new transformation rules have an inverse role with respect to the previous ones.

In Sect. 4.2, the complete reasoning process is described by the definition of a Procedure $\mathcal{R}$, and of two supporting Algorithms $\mathcal{A}_1$ and $\mathcal{A}_2$, with related validity theorems.

Given the outputs of Procedure $\mathcal{P}$, first, all the quantified formulas are deleted from the sequents of S, thus obtaining a set of open sequents. The construction rules are then applied to them, thus obtaining a pair of sets of A-clauses and of S-clauses. These transformation steps are accomplished by Algorithm $\mathcal{A}_1$.

Next, two families of sets of A-clauses and of S-clauses are built, by making all the combinations of the clauses generated by Algorithm $\mathcal{A}_1$. Two new sets of A-clauses and of S-clauses are obtained, by connecting the elements of the sets of the two families. These two steps are accomplished by Algorithm $\mathcal{A}_2$.

In further steps of $\mathcal{R}$, each occurrence of every predicate symbol is substituted by a quantified formula constructed according to the current values of Θ and Y' (output of $\mathcal{P}$). Then, in the final step, two sets of formulas in normal forms are built.

4.1 Construction rules

Since the aim is to build formulas in conjunctive or disjunctive form, only two connectives (namely $\neg, \wedge$ or $\neg, \vee$) need to be considered in the construction rules. Hence the two sets of construction rules are $\mathbf{R}^2_{\text{gen}} = \{\underline{\mathbf{S}\neg}, \underline{\mathbf{A}\wedge}\}$ and $\mathbf{R}^2_{\text{abd}} = \{\underline{\mathbf{A}\neg}, \underline{\mathbf{S}\vee}\}$, where:

$\underline{\mathbf{S}\neg}$: from the sequent $\Gamma \Rightarrow \Delta_1, \sigma, \Delta_2$ the sequent

$$\Gamma, \neg\sigma \Rightarrow \Delta_1, \Delta_2$$

is derived;

<u>A¬</u>: from the sequent $\Gamma_1, \alpha, \Gamma_2 \Rightarrow \Delta$ the sequent

$$\Gamma_1, \Gamma_2 \Rightarrow \neg\alpha, \Delta$$

is derived;

<u>A∧</u>: from the sequent $\Gamma_1, \alpha_1, \alpha_2, \Gamma_2 \Rightarrow \Delta$ the sequent

$$\Gamma_1, \alpha_1 \wedge \alpha_2, \Gamma_2 \Rightarrow \Delta$$

is derived;

<u>S∨</u>: from the sequent $\Gamma \Rightarrow \Delta_1, \sigma_1, \sigma_2, \Delta_2$ the sequent

$$\Gamma \Rightarrow \Delta_1, \sigma_1 \vee \sigma_2, \Delta_2$$

is derived.

These four rules can be proved to transform valid sequents into valid sequents on the basis of Property (xi).

4.2 Construction of a valid sequent

In the following algorithms, steps are numbered with pairs of integers n and n' to stress the duality of the process: Step n always refers to generation while Step n' refers to abduction.

The Algorithm $\mathcal{A}_1$ generates a pair of sets of A-clauses and of S-clauses, given the set S (i.e., the output of Procedure $\mathcal{P}$) as input:

Algorithm $\mathcal{A}_1$
Input: a set S of atomic non valid sequents;
Output: a pair of sets $\langle \Sigma^A, \Sigma^S \rangle$, of A-clauses and of S-clauses, respectively.
Step 1: *remove* all the quantified formulas from the sequents of S;
Step 2: repeat
    apply rule <u>S¬</u> to each sequent of S
  until every such a sequent has an empty succedent;
  let S_1^A be the resulting set of A-open sequents;
Step 2′: repeat
    apply rule <u>A¬</u> to each sequent of S
  until every such a sequent has an empty antecedent;
  let S_1^S be the resulting set of S-open sequents;
Step 3: repeat
    apply rule <u>A∧</u> to each A-open sequent of S_1^A
  until every such a sequent has only one wff (in the form of conjunction of literals) in the antecedent;
  let S_2^A be the resulting set of A-open sequents;
Step 3′: repeat
    apply rule <u>S∨</u> to each S-open sequent of S_1^S

until every such a sequent has only one wff (in the form of disjunction of literals) in the succedent;
let S_2^S be the resulting set of S-open sequents;
Step 4: *build* the set Σ^A of all the antecedents of the A-open sequents of S_2^A;
Step 4′: *build* the set Σ^S of all the succedents of the S-open sequents of S_2^S;
Step 5: *let* $\langle \Sigma^A, \Sigma^S \rangle$ be the resulting pair of sets.

The following theorems can be proved for Algorithm $\mathcal{A}_1$.

Theorem 3. Let S and $\langle \Sigma^A, \Sigma^S \rangle$ be the input and output of Algorithm $\mathcal{A}_1$. Let τ be a wff. For any sequent $\Gamma \Rightarrow \Delta \in S$ there exists $\alpha \in \Sigma^A$ such that the sequent $\Gamma \Rightarrow \Delta, \tau$ is valid iff the sequent $\alpha \Rightarrow \tau$ is valid.

Proof. Let $\Gamma \Rightarrow \Delta$ be an atomic non valid sequent of S, and τ a wff. Let $\Gamma \Rightarrow \Delta, \tau$ be a valid sequent. The execution of Steps 1, 2, and 3 of Algorithm $\mathcal{A}_1$ on all the component formulas of the sequent $\Gamma \Rightarrow \Delta, \tau$, but τ, produces the valid sequent $\alpha \Rightarrow \tau$ (given the soundness of the construction rules).

Let α be an A-clause of Σ^A, obtained by the execution of Steps 1, 2, and 3 of Algorithm $\mathcal{A}_1$ on a sequent $\Gamma \Rightarrow \Delta$ of S, and τ be a wff. Let $\alpha \Rightarrow \tau$ be a valid sequent, then the application of the decomposition rules to only its antecedent produces the valid sequent $\Gamma' \Rightarrow \Delta', \tau$. The sequent $\Gamma \Rightarrow \Delta, \tau$ is also valid, because it is either equal to or it can be obtained from the sequent $\Gamma' \Rightarrow \Delta', \tau$ by introducing (in the antecedent or in the succedent) the quantified formulas removed in Step 1 of Algorithm $\mathcal{A}_1$ from (the antecedent or the succedent of) the sequent $\Gamma \Rightarrow \Delta$ (given the Properties (viii) and (ix)). □

Theorem 3′. Let S and $\langle \Sigma^A, \Sigma^S \rangle$ be the input and output of Algorithm $\mathcal{A}_1$. Let β be a wff. For any sequent $\Gamma \Rightarrow \Delta \in S$ there exists $\sigma \in \Sigma^S$ such that the sequent $\beta, \Gamma \Rightarrow \Delta$ is valid iff the sequent $\beta \Rightarrow \sigma$ is valid.

The proof is analogous to that of Theorem 3.

Algorithm $\mathcal{A}_2$
Input: a pair $\langle \Sigma^A, \Sigma^S \rangle$ of sets of A-clauses and S-clauses;
Output: a pair $\langle \Sigma_D^A, \Sigma_C^S \rangle$ of sets of disjunctions of A-clauses and of conjunctions of S-clauses, respectively.
Step 1: for each A-clause $\alpha_i = \alpha_{i1} \wedge \alpha_{i2} \wedge \ldots \wedge \alpha_{in_i}$ of Σ^A do
build the set A_i^A of all the A-clauses whose elements are simple combinations of rank k, for $k = 1, \ldots, n_i$ of the set $\{\alpha_{i1}, \alpha_{i2}, \ldots, \alpha_{in_i}\}$;
let $A^A = \{A_1^A, \ldots, A_n^A\}$ be the resulting set;
Step 1′: *build* the set A^S similarly to Step 1;
Step 2: *build* the set $\Sigma_D^A = \{\tau_1, \ldots, \tau_m\}$ of all the possible disjunctions $\tau_i = \tau_{i1} \vee \ldots \vee \tau_{in}$ of A-clauses, such that $\tau_{ij} \in A_j^A$, for $j = 1, \ldots, n$.
Step 2′: *build* the set Σ_C^S similarly to Step 2;
Step 3: *let* $\langle \Sigma_D^A, \Sigma_C^S \rangle$ be the resulting pair of sets.

The following theorems can be proved for Algorithm $\mathcal{A}_2$.

Theorem 4. Let $\langle \Sigma^A, \Sigma^S \rangle$ and $\langle \Sigma^A_D, \Sigma^S_C \rangle$ be the input and the output of Algorithm $\mathcal{A}_2$, where $\Sigma^A = \{\alpha_1, \ldots, \alpha_n\}$. For any $\tau \in \Sigma^A_D$ the sequents $\alpha_i \Rightarrow \tau$, for $i = 1, \ldots, n$, are valid.

Proof. Let $\alpha_i \in \Sigma^A$ be the A-clause $\alpha_{i1} \wedge \ldots \wedge \alpha_{in_i}$. Let $\tau \in \Sigma^A_D$ be the formula $\tau_1 \vee \ldots \vee \tau_n$ (disjunction of A-clauses), where $\tau_i = \alpha_{ik_1} \wedge \ldots \wedge \alpha_{ik_l}$, for $i = 1, \ldots, n$ with $1 \leq k_j \leq n_i$, for $j = 1, \ldots, l$, by construction. The sequent $\alpha_i \Rightarrow \tau$ can be rewritten as

$$\alpha_{i1} \wedge \ldots \wedge \alpha_{in_i} \Rightarrow \tau_1 \vee \ldots \vee \tau_i \vee \ldots \vee \tau_n \ ,$$

and also as

$$\alpha_{i1} \wedge \ldots \wedge \alpha_{ik_1} \wedge \ldots \wedge \alpha_{ik_l} \wedge \ldots \wedge \alpha_{in_i} \Rightarrow \tau_1 \vee \ldots \vee (\alpha_{ik_1} \wedge \ldots \wedge \alpha_{ik_l}) \vee \ldots \vee \tau_n \ .$$

By an appropriate application of rule $\underline{\wedge \mathbf{A}}$ and of rule $\underline{\vee \mathbf{S}}$, the following basic valid sequent can be derived

$$\alpha_{i1}, \ldots, \alpha_{ik_1} \wedge \ldots \wedge \alpha_{ik_l}, \ldots, \alpha_{in_i} \Rightarrow \tau_1, \ldots, \alpha_{ik_1} \wedge \ldots \wedge \alpha_{ik_l}, \ldots, \tau_n \ . \quad \square$$

Theorem 4′. Let $\langle \Sigma^A, \Sigma^S \rangle$ and $\langle \Sigma^A_D, \Sigma^S_C \rangle$ be the input and the output of Algorithm $\mathcal{A}_2$, where $\Sigma^S = \{\sigma_1, \ldots, \sigma_m\}$. For any $\beta \in \Sigma^S_C$ the sequents $\beta \Rightarrow \sigma_j$, for $j = 1, \ldots, m$, are valid.

The proof is analogous to that of Theorem 3.

Procedure $\mathcal{R}$ for the complete reasoning process follows. In the sequel, the expression $P_i[x_{i_1}, \ldots, x_{i_{n_i}}]$ is used to denote the i-th occurrence of the predicate symbol P on terms $t_1, \ldots, t_n$, where $x_{i_1}, \ldots, x_{i_{n_i}}$ are all (and only) the symbols occurring in them.

Procedure $\mathcal{R}$
Input: a non valid sequent $\Gamma \Rightarrow \Delta$; the non empty sets S, Θ and Y' output of Procedure $\mathcal{P}$ with $\Gamma \Rightarrow \Delta$ as input;
Output: a pair $\langle \Sigma^A_{DNF}, \Sigma^S_{CNF} \rangle$ of sets of first order wffs in prenex DNF and prenex CNF.
Step 1: *apply* Algorithm $\mathcal{A}_1$ to S;
Step 2: *apply* Algorithm $\mathcal{A}_2$ to the pair $\langle \Sigma^A, \Sigma^S \rangle$, output of Step 1, thus obtaining the pair of sets $\langle \Sigma^A_D, \Sigma^S_C \rangle$;
let $Y'' = \{y_0, \ldots, y_n\}$, with $n \geq 0$, be the set of the first elements of the pairs of Θ;
let $Z = \{z_1, z_2, \ldots\}$ be a set of symbols not occurring in Y'';

Step 3: if all the pairs in Θ have an empty set as second component (i.e., the wffs of $\Gamma \Rightarrow \Delta$ are all in CP_0), then *let* Σ^A_{DNF} be Σ^A_{D} and Σ^S_{CNF} be Σ^S_{C} and *perform* Step 6, else *perform* Steps 4 and 5, for the generation process, and Steps 4′ and 5′, for the abduction process;

Step 4: for each occurrence of every predicate symbol P in the disjunction of A-clauses of Σ^A_{D} do

let $P_i[y_{i_1}, \ldots, y_{i_j}, \ldots y_{i_{m_i}}]$, with $m_i \geq 1$, be the i-th occurrence of the predicate symbol P;

if $y_{i_j} \in Y'$, then *substitute* the formula $\forall z_k P_i[y_{i_1}, \ldots, z_k, \ldots y_{i_{m_i}}]$ for the i-th occurrence of the predicate symbol P in Σ^A_{D}, else

if $y_{i_j} = y_k$, with $y_k \in Y''$, for all i (i.e., for all the occurrences of P), then *substitute* the formula $\exists z_k P_i[y_{i_1}, \ldots, z_k, \ldots y_{i_{m_i}}]$ for the occurrences of P in Σ^A_{D}, else *substitute* the formula $\forall z_k P_i[y_{i_1}, \ldots, z_k, \ldots y_{i_{m_i}}]$ for the occurrences of P in Σ^A_{D};

Step 4′: *modify* the set Σ^S_{C} similarly to Step 4, by exchanging the role of the existential and universal quantifiers;

Step 5: for each occurrence of formulas of Σ^A_{D} including both existential and universal quantifiers do

interchange the order of the quantifiers according to the order of the instantiations of quantified formulas, as included in the second elements of the pairs of Θ (i.e., the order of the variables that have been instantiated);

normalize the formulas of Σ^A_{D} by applying the laws of CP_1 and *let* Σ^A_{DNF} be the resulting set;

Step 5′: *build* the set Σ^S_{CNF} similarly to Step 5;

Step 6: *let* $\langle \Sigma^A_{\mathrm{DNF}}, \Sigma^S_{\mathrm{CNF}} \rangle$ be the resulting pair of sets.

The following theorems can be proved for Procedure $\mathcal{R}$.

Theorem 5. Let $\Gamma \Rightarrow \Delta$ and $\langle \Sigma^A_{\mathrm{DNF}}, \Sigma^S_{\mathrm{CNF}} \rangle$ be the input and the output of $\mathcal{R}$. For all $\sigma \in \Sigma^A_{\mathrm{DNF}}$ the sequent $\Gamma \Rightarrow \Delta, \sigma$ is valid.

Proof. Let $\Gamma \Rightarrow \Delta$ be a non valid sequent. Let $S = \{\Gamma_1 \Rightarrow \Delta_1, \ldots, \Gamma_n \Rightarrow \Delta_n\}$ be the non empty output of $\mathcal{P}$ applied to the sequent $\Gamma \Rightarrow \Delta$. Let $\langle \Sigma^A_{\mathrm{D}}, \Sigma^S_{\mathrm{C}} \rangle$ be the output of $\mathcal{A}_2$, as in Step 2 of $\mathcal{R}$. For any $\tau \in \Sigma^A_{\mathrm{D}}$ $\Gamma \Rightarrow \Delta, \tau$ is valid, by Theorems 2–4.

For any $\tau \in \Sigma^A_{\mathrm{D}}$, Steps 3–5 of $\mathcal{R}$ generate a (possibly quantified) wff σ (in disjunctive normal form), according to the current values of Θ and Y' (output of $\mathcal{P}$). Then, the sequent $\Gamma \Rightarrow \Delta, \sigma$ is valid, since the execution of $\mathcal{P}$, with this sequent as input, terminates with an empty output S. Indeed, this execution of $\mathcal{P}$ generates the sequents $\Gamma_i \Rightarrow \Delta_i, \tau, \sigma_i$, for $i = 1, \ldots, n$ (where σ_i are either empty or quantified formulas), that are valid by Theorem 2 and by Property (viii). □

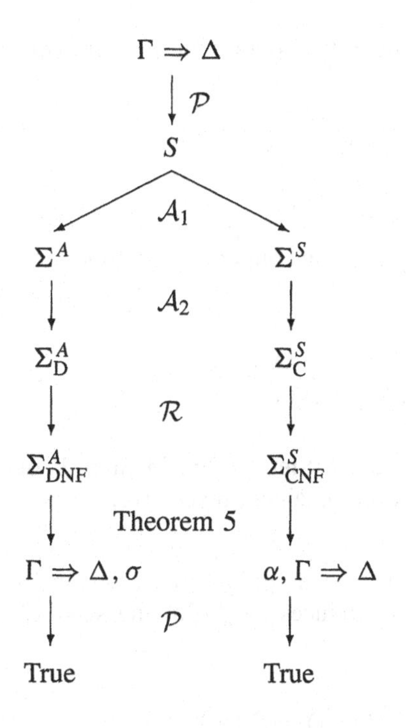

Fig. 2. Complete reasoning process shown in tree form, where vertical arrows indicate the derivation of sets of sequents and of formulas by the application of the procedures and the algorithms labelling the arrows; valid sequents are indicated by the constant "true"

Theorem 5′. Let $\Gamma \Rightarrow \Delta$ and $\langle \Sigma^A_{\text{DNF}}, \Sigma^S_{\text{CNF}} \rangle$ be the input and the output of $\mathcal{R}$. For all $\alpha \in \Sigma^S_{\text{CNF}}$, the sequent $\alpha, \Gamma \Rightarrow \Delta$ is valid.

The proof is analogous to that of Theorem 5.

The complete reasoning process defined in this section is shown in its tree form in Fig. 2. In particular, the duality of the processes of generation and of abduction clearly appears in the tree.

4.3 An example of construction of a valid sequent

An example of composition of a valid sequent in a generation process follows. In this example, the set of the following two sequents (output of Procedure $\mathcal{P}$ in the example of Sect. 3.1) is assumed as input to Algorithm $\mathcal{A}_1$.

(p_1) $\quad \forall x_1(P_1(x_1) \supset \neg P_2(x_1)), P_2(y_1), P_3(y_1) \Rightarrow P_1(y_0), P_1(y_1)$

(p_2) $\quad \forall x_1(P_1(x_1) \supset \neg P_2(x_1)), P_2(y_1), P_3(y_1) \Rightarrow P_2(y_0), P_1(y_1)$

Step 1 of Algorithm $\mathcal{A}_1$ produces the following open sequents:

(q_1) $\quad P_2(y_1), P_3(y_1) \Rightarrow P_1(y_0), P_1(y_1)$

(q_2) $\quad P_2(y_1), P_3(y_1) \Rightarrow P_2(y_0), P_1(y_1)$

In Step 2 of Algorithm $\mathcal{A}_1$ the rule $\underline{\mathbf{S}\neg}$ is applied, thus obtaining the following A-open sequents:

$$(r_1) \qquad \neg P_1(y_0), \neg P_1(y_1), P_2(y_1), P_3(y_1) \Rightarrow$$
$$(r_2) \qquad \neg P_1(y_1), \neg P_2(y_0), P_2(y_1), P_3(y_1) \Rightarrow$$

In Step 3 of Algorithm $\mathcal{A}_1$ the rule $\underline{\mathbf{A}\wedge}$ is applied, thus obtaining the following A-clauses:

$$(\alpha_1) \qquad \neg P_1(y_0) \wedge \neg P_1(y_1) \wedge P_2(y_1) \wedge P_3(y_1)$$
$$(\alpha_2) \qquad \neg P_1(y_1) \wedge \neg P_2(y_0) \wedge P_2(y_1) \wedge P_3(y_1)$$

Similarly, by applying the rules $\underline{\mathbf{A}\neg}$ and $\underline{\mathbf{S}\vee}$ as in Step 2′ and in Step 3′ of Algorithm $\mathcal{A}_1$, S-open sequents and S-clauses are obtained, respectively.

Next, the execution of Step 1 of Algorithm $\mathcal{A}_2$, when applied to the set of A-clauses $\{\alpha_1, \alpha_2\}$ (generated by Algorithm $\mathcal{A}_1$) produces the following sets A_1^A and A_2^A of clauses:

$$\begin{aligned} A_1^A = \{&\neg P_1(y_1), \neg P_2(y_0), P_2(y_1), P_3(y_1), \neg P_1(y_1) \\ &\wedge \neg P_2(y_0), \ldots, \neg P_1(y_1) \wedge \neg P_2(y_0) \wedge P_2(y_1)\} \\ A_2^A = \{&\neg P_1(y_0), \neg P_1(y_1), \ldots, \neg P_1(y_0) \wedge P_1(y_1), \ldots\} \end{aligned}$$

Then, Step 2 of Algorithm $\mathcal{A}_2$, when applied to these two sets, produces the following set of clauses:

$$\Sigma^A = \{\neg P_1(y_1) \vee \neg P_1(y_0), \neg P_1(y_1) \vee \neg P_1(y_1), \ldots\} .$$

By considering the current values of Θ and Y' (as obtained by $\mathcal{P}$ in the example of Sect. 3.1) and by observing that $\neg P_1(y_1) \wedge P_3(y_1) \in A_1^A \cap A_2^A$, we conclude that $\neg P_1(y_1) \wedge P_3(y_1) \in \Sigma^A$. Furthermore, the symbol y_1 has been used in the instantiation of an existential quantified formula, then the following clause is obtained by the application of $\mathcal{R}$:

$$\exists z(\neg P_1(z) \wedge P_3(z)) .$$

Then, by Theorem 5, the Procedure $\mathcal{P}$ with

$$\forall x_1(P_1(x_1) \supset \neg P_2(x_1)), \exists x_2(P_2(x_2) \wedge P_3(x_2)) \Rightarrow \exists z(P_3(z) \wedge \neg P_1(z))$$

as input, terminates with an empty output set S, and the sequent obtained by the composition process is valid.

5 Conclusions

In this paper, a sequent calculus for automated reasoning has been presented. It is an extension of the sequent calculus to deal with generative and abductive problems, as well as with verificative problems. The proposed method solves generative and abductive logic problems for a non valid sequent by building sets of wffs that, when added to the given sequent, make it valid. These sets include all the wffs derivable from the given sequent by applying a given set of transformation rules.

A uniform methodological and computational approach has been taken in the definition of the proposed sequent calculus. This approach is particularly suitable to support the full automation of the reasoning method.

The reasoning process of building wffs, for generation and abduction, follows symmetric computational steps (cf. Fig. 2). The analogy of generation and abduction is stressed by noticing that the two sets of wffs produced by the reasoning process are in a one-to-one correspondence, given the duality of the construction. In this context, other properties of these sets of wffs are under investigation (e.g., to provide criteria for choosing a wff according to particular characteristics of a given problem and for improving the efficiency of the method) together with careful analysis of the relationship between deduction and abduction.

The presentation of results refers to first-order logic, so as to implicitly include the case of propositional logic (in which Procedure $\mathcal{P}$ terminates). Further extensions of the method (e.g., to deal with equality) can be obtained by only enlarging the set of transformation rules. As far as the automation of the proposed method is concerned, a prototype implementation has been developed within an object-oriented programming paradigm. In this implementation, the automation of the reasoning process for CP_1 has been obtained as an extension of that for CP_0, on the basis of the mechanisms of inheritance (Bonamico and Cioni 1988, Bonamico et al. 1990). For the Procedure $\mathcal{P}$ a halting criterium (based on the length of computation) is defined, to overcome non-terminating cases.

Moreover, an interactive access allows the user to monitor the execution, to (possibly) modify the selection strategy embedded in $\mathcal{P}$ and to stop long computations. In particular, one can stop the execution of $\mathcal{P}$ as soon as an atomic non valid sequent has been introduced into the set S and proceed with further steps of procedure $\mathcal{R}$ to build only some of the possible solutions to generative and abductive problems.

The prototype is also a basic module of a software system for symbolic mathematics allowing for both computing and reasoning on mathematical objects for which full axiomatization is provided (Limongelli and Temperini 1992). Actually, there is a strong request for extending systems for symbolic mathematics by theorem proving capabilities, a subject whose evolution requires an intimate interaction of mathematics and artificial intelligence (Miola 1991). This is also a basic research area in the design and implementation of symbolic computation systems of a new generation (Miola 1990, 1995).

A very interesting application of the proposed sequent calculus is in the design of intelligent tutoring systems (ITS). As presented by Aiello et al. (1992), the expert module of an ITS can employ the proposed method as an automated deduction method to allow the student to either verify the validity of a given theorem or discover the conditions under which the theorem is valid. In this application, an interactive step-by-step execution of the reasoning procedure well supports the tutoring activity of the system.

Further applications of the proposed method are under investigation, with particular attention to diagnosis problems where generation and abduction techniques play a central role.

Acknowledgement

Work partially supported by CNR under the project "Sistemi Informatici e Calcolo Parallelo".

References

Aiello, L., Colagrossi, A., Micarelli, A., Miola, A. (1992): Building the expert module for ITS in mathematics: a general reasoning apparatus. In: Nwana, H. S. (ed.): Mathematical intelligent learning environments. Intellectic Publishing, Oxford, pp. 35–51.

Bibel, W. (1987): Automated theorem proving. F. Vieweg und Sohn, Wiesbaden.

Bonamico, S., Cioni, G. (1988): Embedding flexible control strategies into object oriented languages. In: Mora, T. (ed.): Applied algebra, algebraic algorithms and error-correcting codes. Springer, Berlin Heidelberg New York Tokyo, pp. 454–457 (Lecture notes in computer science, vol. 357).

Bonamico, S., Cioni, G., Colagrossi, A. (1990): A gentzen based deduction method for mathematical problems solving. In: Balagurusamy, E., Sushila, B. (eds.): Computer systems and applications: recent trends. Tata McGraw-Hill, New Delhi, pp. 102–113.

Ebbinghaus, H. D., Flum, J., Thomas, W. (1984): Mathematical logic. Springer, New York Berlin Heidelberg.

Gallier, J. H. (1986): Logic for computer science. Harper and Row, New York.

Girard, J., Taylor, P., Lafont, Y. (1989): Proofs and types. Cambridge University Press, Cambridge.

Hermes, H. (1973): Introduction to mathematical logic. Springer, Berlin.

Kowalski, R. (1979): Logic for problem solving. North-Holland, New York.

Limongelli, C., Temperini, M. (1992): Abstract specification of structures and methods in symbolic mathematical computation. Theor. Comput. Sci. 104: 89–107.

Miola, A. (ed.) (1990): Design and implementation of symbolic computation systems. Springer, Berlin Heidelberg New York Tokyo (Lecture notes in computer science, vol. 429).

Miola, A. (1991): Symbolic computation and artificial intelligence. In: Jorrand, P., Kelemen, J. (eds.): Fundamentals of Artificial Intelligence Research, FAIR'91. Springer, Berlin Heidelberg New York Tokyo, pp. 244–255 (Lecture notes in computer science, vol. 535).

Miola, A. (ed.) (1995): Design and implementation of symbolic computation systems. J. Symb. Comput. 19.

A sequent calculus machine for symbolic computation systems

P. Bertoli, G. Cioni, A. Colagrossi, and P. Terlizzi

1 Introduction

In literature many proposals for mathematical problem solving can be found, those allowing for manipulating the properties of the mathematical objects involved. If these properties have been expressed in first-order logic language, this task can be performed by using automated deduction methods developed in the frame of natural deduction (see, e.g., Beeson 1989, Suppes and Takahashi 1989) with the advantage of a friendly use. In this frame a method for reasoning on properties, based on a sequent calculus, has been developed and proposed in Cioni et al. (1997, 1995); its characteristics, that make it well suited to be embedded in a symbolic mathematical computation system, are also stressed in those papers, also considering different kinds of problems to be approached. In fact, such a sequent calculus is a single method for performing extended deduction, i.e., for solving verificative, generative, and abductive problems in one methodological and integrated way.

This paper presents a system which implements such a sequent calculus and refers to Cioni et al. (1995) for the technical characteristics (properties, theorems, algorithms, and procedures) of the method. This sequent calculus system has been implemented and embedded in an object-oriented programming environment. The system allows one to operate in two different ways: interactively or in a fully automatized mode. This characteristic refers, e.g., to the choice of strategies for the selection of the deduction rules which is left to the user in the case of the interactive mode and is made by the system in the fully automatized mode.

The system requires as input a sequent and a list of execution parameters (such as, e.g., the maximal depth of the deduction tree that the system will build); then the verificative process starts with the selected mode of operating. The output of the verificative procedure is one of the following three messages:

1. the given input sequent is valid;
2. the given input sequent is not valid;
3. nothing can be said about the given input sequent: modify the execution parameters and retry.

In the case of a not valid sequent, the system outputs also a set of atomic sequents (i.e., sequents which cannot be further decomposed).

If the given sequent is not valid, then generative and/or abductive procedures can be activated, taking as input the set of atomic sequents, output of the verificative procedure. Also these procedures can be executed in both interactive and fully automatized modes of operating.

The chosen object-oriented methodology is well suited to the design of the proposed system, given the hierarchical structure of the defined sequent calculus whose kernel is, of course, the propositional calculus.

The system is carried out in CLOS on a SUN machine (Cioni and Miola 1992). In what follows the object-oriented design of the system is first analyzed, together with its benefits, then the design and the adopted implementation principles are described in some details. In particular, the characteristics of CLOS and the structure of the software system are discussed. The last section of this paper presents some working sessions with the described system.

2 Object-oriented design

The main feature to be offered by an automated deduction system embedded in a programming environment is the capability to manipulate properties of mathematical objects using inference rules with the same methodological approach followed during the computational steps. For this reason an object-oriented design has been followed, taking also into consideration the peculiarities of mathematical objects, such as their axiomatical structure and their high abstraction level.

Moreover, an object-oriented design allows an easy extendibility of the system which needs to be frequently updated on the basis of new results of research work yet in progress.

The two major benefits of this design are the naturalness in the definition of mathematical objects and the correctness which can be guaranteed at every step of the execution. In fact, the objects have been considered both as computable entities, manipulated by appropriate numeric and symbolic methods, and as entities owing properties, expressed in a logical formalism. On these properties different inference steps have to be performed. These strictly connected and uniformly operating two factors allow for verification of the correctness of all the elements of the system both during the compilation and during the execution.

In regard to the structure of the system, let us note that the sequent calculus is a structured method and can be easily represented following an object-oriented approach. Then, on top of a core devoted to propositional logic calculus and constituted by classes and methods for applying the decomposition rules related to connectives, predicative calculus has been implemented by adding other classes and methods concerning quantifiers and by using inheritance relationships among classes. So methods did not have to be completely redefined, but only specialized. Finally, a complete sequent calculus, with methods for proving and reasoning, terminates this modular construction.

In our implementation a functional object-oriented programming language has been chosen, namely CLOS (common Lisp object system) (Paepcke 1993). The basic difference between CLOS and other object-oriented languages is that methods are not encapsulated in the class structure, but simply collected in some generic functions, so implementing polymorphism in a way that combines object-oriented style with functional style. We took advantage of this feature of CLOS in our implementation. In fact, from a general point of view the removal of encapsulation constraints guarantees, when operating without data protection, flexibility in the objects design and a greater modularity allowing the user of a generic function to ignore the implementation details of the collected methods. Moreover, we exploited the possibility offered by generic functions to execute code belonging to different classes linked in a hierarchical structure. The user can define own methods combination, as a mechanism of choosing the method to be executed at the reference of a method call, where the actual method can be also the execution of logically linked methods. So, in the implementation of inference rules we have linked in a hierarchical structure methods performing the decomposition, by deriving classes representing more specific mechanisms from classes representing more general mechanisms.

Presently the integration of computational services into the environment is in progress, with the purpose of reasoning in given predefined interpretation domains. That means to approach the problem of verifying the satisfiability of a given set of properties which refer to specific objects belonging to axiomatizable mathematical theories.

In order to verify satisfiability, computational methods of the considered mathematical theory have to be applied. Then, either these methods are yet present in the computational system, or they have to be supplied by the user when needed. In the latter case the choice of CLOS as implementation language proves very useful because the functional programming style, together with the object-oriented paradigm, perfectly fits with this approach, allowing the treatment of functions as data.

3 Design and implementation remarks

In this section we present a general view of the system design, following a top-down approach in the refinement process. The basic program structure contains a first step devoted to the syntactical control, the setting of the system by the user, the deduction step and, possibly, a validation step. In fact, three alternative results can be obtained at the end of the deduction process: a validity demonstration for the given sequent (i.e., a demonstration tree whose leaves represent axioms), a non-validity demonstration (i.e., a counter-example tree, with at least one non axiom leaf), or the abort of the deduction process (due to the limits which have been fixed on terms of universe dimension and of tree depth, necessary for the intrinsic semidecidibility of first-order logic). In the second case, a validation process can be activated, generating (or abducing) formulas that, once added to the thesis (or to the hyphothesis) of the given sequent, allow to obtain a valid sequent. The basic program structure is cyclical, allowing consecutive theorem proofs.

Following the object-oriented programming paradigm, we represent a formula by defining a class *Formula* with an internal string representation. Then, we introduce a class *Set-form*, representing a set of formulas with a suitable representation for antecedent and succedent of the sequent. Finally, a class *Sequent*, with two instance variables representing antecedent and succedent as formula

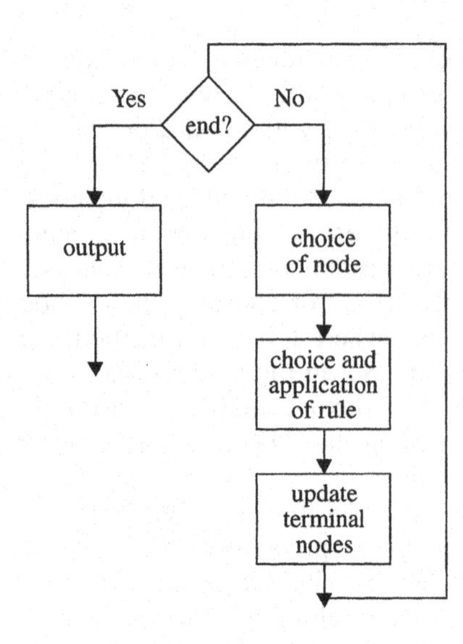

Fig. 1. General structure of the system

sets and a third instance variable that is a set of one or two children sequents (derived by a rule application) is defined. In this way, a deduction tree is completely represented starting from the root sequent.

During the step devoted to setting the system, the user's choices are specified: the tree development strategy (breadth, or depth, or interactive); the selection mode (automatic or interactive) for the decomposition rules to be applied in a given state of the process; the space limits in the deduction process, to avoid non-termination situations.

In regard to the deduction block, that is the program core, let us note that this block is devoted to build the deduction tree in order to validate the given sequent.

First refinement (Fig. 1) shows, after a test for stopping the process under user request, the main cycle of the application of the decomposition rules. The algorithm uses a terminal nodes list, from which a node to be developed is chosen on the basis of the current strategy. A suitable decomposition rule is then applied to this node, and the terminal nodes list is updated. Let us remark that we refer to nodes and not to sequents: in fact, the generation of the deduction tree follows a general algorithm of building a generic binary tree. So, we derive sequents from nodes, in an inheritance relationship. In this way, a deduction tree is a tree whose nodes have a sequent as mark and for which a specific decomposition rule mechanism, in order to derive one node from another, exists.

The choice and application rule block is expanded as shown in Fig. 2. All possible rules are applied to the actual node, according to a predefined priority order (based upon heuristic considerations whose aim is to minimize the dimensions of the tree). Then, on the basis of the current strategy, a rule is selected and the decomposition made on the actual node increases the tree. Let us remark that, in automatic mode, a faster deduction process can be performed avoiding the application of all possible rules on the current node and choosing and applying the first one applicable following a predefined priority order.

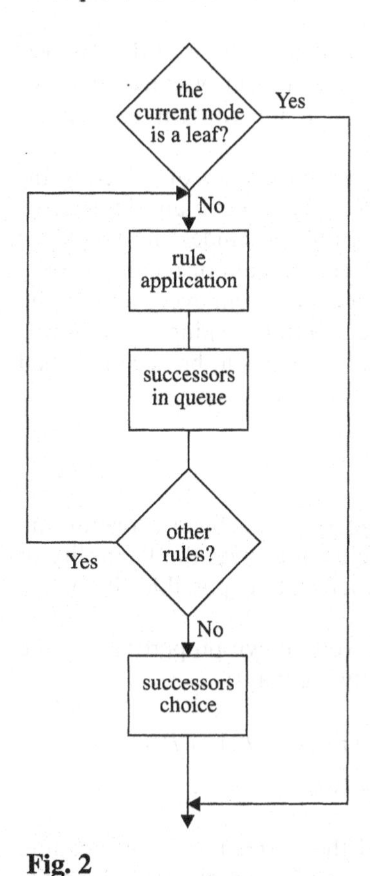

Fig. 2

Fig. 3

Fig. 2. Choice and application rules block
Fig. 3. Validation block

The output block has the purpose of tracing the deduction process, once it has been terminated. This feature is very important in automatic mode when the user cannot follow the demonstration step-by-step. The available data are: the complete demonstration, printable to video or file; information about computational complexity (CPU time, nodes number, and tree depth); and, finally, for each node a complete description and history (father node, applicable rules, choosen rule, etc.).

The validation block can be decomposed as depicted in Fig. 3. The validation process is performed, starting from the tree of a non valid sequent, through the algorithm presented in Cioni et al. (1995). The block for optimizing leaves includes the elimination from the tree of valid or repeated leaves, the following removal of quantified formulas from remaining leaves of the tree and the application of recomposition rules. In this way, starting from the leaves of the deduction tree whose root is a not valid sequent, we obtain a set of leaves with empty antecedent (or succedent). Let us remark that the recomposition rules used in the validation process are in some sense inverse to those of decomposition used during the verification process. So, in a strict object-oriented style, we use the common parts of the decomposition rules code, inheriting it, and adding specific necessary elements.

The literals choice block of Fig. 3 provides the possibility of selecting interesting literals in the validation process, supplying automatic and interactive strategies. The block which builds validating formulas includes the block for building clauses which implements the combinatorial process of recomposition. All the validating formulas are built by choosing a clause from every leaf. In the interactive way the program supplies one formula at a time, asking to the user if another formula has to be built, if a verification process has to be accomplished and if the abductive/generative process has to be quited.

4 Use of function symbols

The system allows also for dealing with function symbols. Such a feature, no longer to increase the power of the sequent calculus, improves the flexibility of the entire system by expanding the language and then making possible alternative expressions for the same mathematical property.

For example, it will be possible to express a well known property about the transposed matrix multiplication in both the following ways:

$$\forall a, b, c\ \exists h, k, i\ ((P(a, b, c) \wedge T(a, h) \wedge T(b, i) \wedge T(c, k)) \supset P(i, h, k)) \ ,$$
$$\forall a, b, c\ (P(a, b, c) \supset P(ft(b), ft(a), ft(c))) \ .$$

In order to describe the implementation and the utilization of the feature, some details about the theoretical basis and the use of function symbols in the sequent calculus have to be given. This was unessential for the problem of performing extended deduction in our theoretical reference (Cioni et al. 1997).

The main difficulty deriving from the use of function symbols consists in:

- taking into account the recursivity induced by the definition of terms; this fact essentially means that it will be necessary to check for the arity consistency of the predicate and function symbols;
- modifying the extended deduction procedures given in Cioni et al. (1997), given the impact on the generation of the univers of terms.

Here we will only mention the main modifications operated on the proving procedure as given in Cioni et al. (1997). They follow from Gallier's (1986) approach and consist essentially in introducing $n+3$ new lists: *TERMS*, $TERM_0$, and $AVAIL_i$, for $i = 0, \ldots, n$. The elements of *TERMS* are all the terms obtained by applying the functional symbols of the given sequent to the terms (constants and free variables) belonging to the sequent itself. The lists $TERM_0$ and $AVAIL_0$ are initialized to the pair $\langle head(TERMS), \emptyset \rangle$ and $tail(TERMS)$ respectively. The lists $AVAIL_i$ are built up in correspondence to the introduction of n eigenvariables y_i. Their elements are $y_0, \ldots, y_i$, the constants and free variables of the sequent, and all the terms obtained by applying them the functional symbols of the sequent. Again, it is $y_i = head(AVAIL_i)$.

Handling the lists $AVAIL_i$ is a hard task, given the inherent recursivity of the definition of such lists and their non-finite length. The solution devised in this work consists in generating for a given list with an element k only the element $k+1$. In order to do so, an ordering in the generation of the terms of the lists $AVAIL_i$ has been introduced. Such an ordering is obtained considering, for each term generated, the depth of the tree corresponding to the representation of that functional term.

The modifications made to the proving procedure are coherent with the data structures above described.

5 Examples of working sessions

In this section two examples of working sessions are presented. The chosen presentation is very similar to the real output of the system, but some non necessary parts of the dialog are omitted. Actually a more friendly interface is under development in order to use that same presentation tool of the other parts of the TASSO project.

The first working example shows the solution of a verificative problem. The given input sequent is a valid one and is defined using function symbols.

```
    SEQUENT: → VxPx&EyQ(y)>P(fa(u))&EzQ(z)
    Rectified Sequent:
    → Vz1P(z1)&Ez2Q(z2)>P(fa(u))&Ez3Q(z3)
    Starting node: → Vz1P(z1)&Ez2Q(z2)>P(fa(u))&Ez3Q(z3)

    Expanding node(1):
    → Vz1P(z1)&Ez2Q(z2)>P(fa(u))&Ez3Q(z3)
    (level 1)
    Possible successors:

1.  Rule IMPLY succedent (Formula 1):
    Vz1P(z1)&Ez2Q(z2) → P(fa(u))&Ez3Q(z3)

    There is only one rule applicable to the node.
    Result is node (1.1):
    Vz1P(z1)&Ez2Q(z2) → P(fa(u))&Ez3Q(z3)
    (expandible sequent)

    Expanding node(1.1):
    Vz1P(z1)&Ez2Q(z2) → P(fa(u))&Ez3Q(z3)
    (level 2)
    Possible successors:

1.  Rule AND antecedent (Formula 1):
    Vz1P(z1);Ez2Q(z2) → P(fa(u))&Ez3Q(z3)
```

```
2. Rule AND succedent (Formula 1):
   Vz1P(z1)&Ez2Q(z2) → P(fa(u))
   and: Vz1P(z1)&Ez2Q(z2) → Ez3Q(z3)

   Choose the rule to be applied to the node among the proposed by
selecting the relevant number: rule n. 1
   Result is node (1.1.1):
   Vz1P(z1);Ez2Q(z2) → P(fa(u))&Ez3Q(z3)
   (expandible sequent)

   Expanding node(1.1.1):
   Vz1P(z1);Ez2Q(z2) → P(fa(u))&Ez3Q(z3)
   (level 3)
   Possible successors:

1. Rule EXIST antecedent (Formula 2):
   Vz1P(z1);Q(y) → P(fa(u)) & Ez3Q(z3)
2. Rule FOR-ALL antecedent (Formula 1):
   Vz1P(z1); P(fa(u)); P(u); P(fa(fa(u))); Ez2Q(z2) →
   P(fa(u)) & Ez3Q(z3)
3. Rule AND succedent (Formula 1):
   Vz1P(z1); Ez2Q(z2) → P(fa(u))
   and: Vz1P(z1); Ez2Q(z2) → Ez3Q(z3)

   Choose the rule to be applied to the node among the proposed by
selecting the relevant number: rule n. 3
   Result is nodes (1.1.1.1) and (1.1.1.2):
   Vz1P(z1); Ez2Q(z2) → P(fa(u))
   (expandible sequent)
   Vz1P(z1); Ez2Q(z2) → Ez3Q(z3)
   (expandible sequent)

   Expanding node(1.1.1.1):
   Vz1P(z1);Ez2Q(z2) → P(fa(u))
   (level 4)
   Possible successors:

1. Rule EXIST antecedent (Formula 2):
   Vz1P(z1); Q(y) → P(fa(u))
2. Rule FOR-ALL antecedent (Formula 1):
   Vz1P(z1); P(fa(u)); P(u); P(fa(fa(u))); P(fa(fa(fa(u))));
   Ez2Q(z2) → P(fa(u)) & Ez3Q(z3)

   Choose the rule to be applied to the node among the proposed by
selecting the relevant number: rule n. 2
   Result is node (1.1.1.1.1):
   Vz1P(z1); P(fa(u)); P(u); P(fa(fa(u))); P(fa(fa(fa(u))));
```

```
Ez2Q(z2) → P(fa(u)) & Ez3Q(z3)
   (Leaf: valid sequent)

   Expanding node(1.1.1.2):
   Vz1P(z1); Ez2Q(z2) → Ez3Q(z3)
   (level 4)
   Possible successors:

1. Rule EXIST antecedent (Formula 2):
   Vz1P(z1); Q(y) → Ez3Q(z3)
2. Rule FOR-ALL antecedent (Formula 1):
   Vz1P(z1); P(fa(u)); P(u); P(fa(fa(u))); P(fa(fa(fa(u))));
   P(fa(fa(fa(fa(u))))); Ez2Q(z2) → Ez3Q(z3)
3. Rule EXIST succedent (Formula 1):
   Vz1P(z1); Ez2Q(z2) → Ez3Q(z3); Q(fa(u)); Q(u);
   Q(fa(fa(u))); Q(fa(fa(fa(u)))); Q(fa(fa(fa(fa(u)))))

   Choose the rule to be applied to the node among the proposed by
selecting the relevant number: rule n. 2
   Result is node (1.1.1.2.1):
   Vz1P(z1); P(fa(u)); P(u); P(fa(fa(fa(u))));
P(fa(fa(fa(fa(u))))); Ez2Q(z2) → Ez3Q(z3)
   (expandible sequent)

   Expanding node(1.1.1.2.1):
   Vz1P(z1); P(fa(u)); P(u); P(fa(fa(u))); P(fa(fa(fa(u))));
P(fa(fa(fa(fa(u))))); Ez2Q(z2) → Ez3Q(z3)
   (level 5)
   Possible successors:

1. Rule EXIST antecedent (Formula 7):
   Vz1P(z1); P(fa(u)); P(u); P(fa(fa(u))); P(fa(fa(fa(u))));
   P(fa(fa(fa(fa(u))))); Q(y) → Ez3Q(z3)
2. Rule EXIST succedent (Formula 1):
   Vz1P(z1); P(fa(u)); P(u); P(fa(fa(u))); P(fa(fa(fa(u))));
   P(fa(fa(fa(fa(u))))); Ez2Q(z2) → Ez3Q(z3); Q(fa(u)); Q(u);
   Q(fa(fa(u))); Q(fa(fa(fa(u)))); Q(fa(fa(fa(fa(u)))))

   Choose the rule to be applied to the node among the proposed by
selecting the relevant number: rule n. 1
   Result is node (1.1.1.2.1.1):
   Vz1P(z1); P(fa(u)); P(u); P(fa(fa(u))); P(fa(fa(fa(u))));
P(fa(fa(fa(fa(u))))); Q(y1) → Ez3Q(z3)
   (expandible sequent)

   Expanding node(1.1.1.2.1.1):
   Vz1P(z1); P(fa(u)); P(u); P(fa(fa(u))); P(fa(fa(fa(u))));
```

```
P(fa(fa(fa(fa(u))))); Q(y1) → Ez3Q(z3)
   (level 6)
   Possible successors:

1. Rule FOR-ALL antecedent (Formula 1):
   Vz1P(z1); P(fa(u)); P(u); P(fa(fa(u))); P(fa(fa(fa(u))));
   P(fa(fa(fa(fa(u))))); P(y1); P(fa(fa(fa(fa(fa(u))))));
   Q(y1) → Ez3Q(z3)
2. Rule EXIST succedent (Formula 1):
   Vz1P(z1); P(fa(u)); P(u); P(fa(fa(u))); P(fa(fa(fa(u))));
   P(fa(fa(fa(fa(u))))); Q(y1) → Ez3Q(z3); Q(fa(u)); Q(u);
   Q(fa(fa(u))); Q(fa(fa(fa(u)))); Q(fa(fa(fa(fa(u))))); Q(y1);
   Q(fa(fa(fa(fa(fa(u))))))

   Choose the rule to be applied to the node among the proposed by
selecting the relevant number: rule n. 2
   Result is node (1.1.1.2.1.1.1):
   Vz1P(z1); P(fa(u)); P(u); P(fa(fa(u))); P(fa(fa(fa(u))));
P(fa(fa(fa(fa(u))))); Q(y1) → Ez3Q(z3); Q(fa(u)); Q(u);
Q(fa(fa(u))); Q(fa(fa(fa(u)))); Q(fa(fa(fa(fa(u))))); Q(y1);
Q(fa(fa(fa(fa(fa(u))))))
   (Leaf: valid sequent)

   Deduction complete; the sequent:
   → VxPx & EyQ(y) > P(fa(u)) & EzQ(z)
   is VALID.
```

The second example has been chosen in order to illustrate the possibility of solving not only verificative problems, but also abductive and generative.

The steps of the working session will proceed as follows:

1. a sequent is given as input to the system;
2. the system performs the verificative step and outputs negative answer together with a set of atomic sequents;
3. the user asks the system to perform a generative process in interactive mode;
4. the system first simplifies the set of atomic sequents, then transforms them in sequents with empty successors;
5. the system asks the user to select at least one literal from each of the simplified and transformed atomic sequents;
6. on the basis of the selected literals the system completes the generative process furnishing a well-formed formula (wff) which, once added to the succedent of the input sequent, makes it valid.

```
SEQUENT:
Vx (P(x) > Q(x)); Ex (Q(x) & R(x) → Ex (P(x) & R(x))
Rectified Sequent:
Vz1 (P(z1) > Q(z1)); Ez2 (Q(z2) & R(z2) → Ez3 (P(z3) & R(z3))
Demonstration in automatic mode
```

```
   .......

   Deduction complete; the sequent
   Vx (P(x) > Q(x)); Ex (Q(x) & R(x) → Ex (P(x) & R(x))
   is NON VALID.

   Do you want to perform a validative process ?
(y, Y/ n, N): y
   Generative or Abductive process?
(g, G/ a, A): g
   Automatic or interactive strategy of choice of literals?
(a, A: i,I): i
   Removing valid sequents and duplicates from leaves ...
   Removing quantified formulas from leaves...
   Applying composition rules NOT succedent...
   ................
   Leaves:
   Node 1.1.1.1.1.1.1.1.2:
Q(y1); R(y1); -P(x1); -P(y1) → /
   Node 1.1.1.1.1.1.1.2.2:
Q(x1); Q(y1); R(y1); -P(x1); -P(y1) → /
   Node 1.1.1.1.1.2.1.1.2:
Q(y1); -R(y1); -R(x1); -P(y1); -P(x1) → /
   Node 1.1.1.1.1.2.1.2.2:
Q(x1); Q(y1); R(y1); -R(x1); -P(y1) → /

   Node 1.1.1.1.1.1.1.1.2:
Antecedent= Q(y1); R(y1); -P(x1); -P(y1)
   Choose at least one literal:
   Q(y1) (y, Y/ n, N): y
   R(y1) (y, Y/ n, N): y
   -P(x1) (y, Y/ n, N): n
   -P(y1) (y, Y/ n, N): n

   Node 1.1.1.1.1.1.1.2.2:
Antecedent= Q(x1); Q(y1); R(y1); -P(x1); -P(y1)
   Choose at least one literal:
   Q(x1) (y, Y/ n, N): n
   Q(y1) (y, Y/ n, N): y
   R(y1) (y, Y/ n, N): y
   -P(x1) (y, Y/ n, N): n
   -P(y1) (y, Y/ n, N): n

   Node 1.1.1.1.1.2.1.1.2:
Antecedent= Q(y1); R(y1); -R(x1); -P(x1); -P(y1)
   Choose at least one literal:
   Q(y1) (y, Y/ n, N): y
```

```
   R(y1) (y, Y/ n, N): y
   -R(x1) (y, Y/ n, N): y
   -P(x1) (y, Y/ n, N): n
   -P(y1) (y, Y/ n, N): n

   Node 1.1.1.1.1.2.1.2.2:
Antecedent= Q(x1); Q(y1); R(y1); -R(x1); -P(y1)
   Choose at least one literal:
   Q(x1) (y, Y/ n, N): n
   Q(y1) (y, Y/ n, N): y
   R(y1) (y, Y/ n, N): y
   -R(x1) (y, Y/ n, N): y
   -P(y1) (y, Y/ n, N): n
   Applying composition rules AND antecedent ...
   ..............
   Validated sequent:
   Vx (P(x) > Q(x)); Ex (Q(x) & R(x)) → Ex (P(x) & R(x);
Vx1 Ey1 ((-R(x1)) v (R(y1)))
```

6 Conclusions and further developments

In this paper we have described the main characteristics of a machine allowing one to perform extended automated deduction in first-order logic by using a sequent calculus.

The characteristics of such a machine include: the methodological approach based on the object-oriented design; a flexible system architecture enabling the adoption of several strategies; the implementation of the software tool carried out in the functional style by using the programming language CLOS; the features offered by the advanced tool, including the basics of the sequent calculus enlarged by enabling the use of the function symbols.

The object-oriented design matches very well with the structure of the logic calculus, considering the propositional calculus as the kernel. The features offered by the machine are the same as described in Cioni et al. (1995), that is they allow one to solve the following three problems:

- verify the validity of a given wff;
- generate a set of wffs;
- abduce a set of wffs.

The available modes of operation are an interactive and a fully automatized. The use of function symbols has also been implemented, so improving the expressiveness of the language.

The machine deserves to be refined in many aspects. Two of the major ones are: to make available the solution of satisfiability problems and implement a user interface adequate to the field of application of the machine. The first aspect involves the integration of computing methods with the extended deduction procedures, enabling the solution of verificative, generative, and abductive problems in a given domain. The second aspect is related to the user point of view, especially in regard to the generative and abductive procedures. The user should be driven by the interface tool through the steps of generation and/or abduction of a wff, offering, moreover, a natural and easy mode of operating, also all those strategies strictly related to the semantical context of the domain considered.

Acknowledgement

Work partially supported by CNR under the project "Sistemi Informatici e Calcolo Parallelo".

References

Beeson, M. J. (1989): Logic and computation in mathpert: an expert system for learning mathematics. In: Kaltofen, E., Watt, S. M. (eds.): Computers and mathematics. Springer, Berlin Heidelberg New York Tokyo, pp. 202–214.

Cioni, G., Colagrossi, A., Miola, A. (1992): A desk-top sequent calculus machine. In: Calmet, J., Campbell, J. A. (eds.): Artificial intelligence and symbolic mathematical computation. Springer, Berlin Heidelberg New York Tokyo, pp. 224–236 (Lecture notes in computer science, vol. 737).

Cioni, G., Colagrossi, A., Miola, A. (1995): A sequent calculus for symbolic computation systems. J. Symb. Comput. 19: 175–199.

Cioni, G., Colagrossi, A., Miola, A. (1997): Deduction and abduction using a sequent calculus. In: Miola, A., Temperini, M. (eds.): Advances in the design of symbolic computation systems. Springer, Wien New York, pp. 198–216 (this volume).

Gallier, J. H. (1986): Logic for computer science. Harper and Row, New York.

Paepcke, A. (1993): Object-oriented programming, the CLOS perspective. MIT Press, Cambridge, MA.

Suppes, P., Takahashi, S. (1989): An interactive calculus theorem-prover for continuity properties. J. Symb. Comput. 7: 573–590.

Automated deduction by connection method in an object-oriented environment

G. Cioni, G. Patrizi, and M. Temperini

1 Introduction

In this paper issues of a logic deduction tool to be integrated in an object-oriented programming (OOP) environment for the manipulation of mathematical objects are presented.

The aim of a symbolic computation system is to provide a set of methods and techniques to support the scientist's work on mathematical objects and models. This work consists in definition, instantiation, and manipulation of mathematical objects. These activities must be supported in numeric computation as well as in symbolic manipulation: straight numeric computations could be performed, but a variety of symbolic treatments should also be supported, e.g., managing exact representation of results and evaluating such results qualitatively.

As a matter of fact, major symbolic computation systems, such as Reduce (Fitch 1985) and Macsyma (Wang and Pavelle 1985), have been successfully used in a variety of applications, in areas of science and engineering. But the fundamental qualitative aspects of the work in mathematics are not well supported, while the analysis of properties of the computed results may be often requested, as part of the whole computation. Moreover, serious lacks of correctness can often occur, depending on the methodological inadequacy of the system design and organization (Miola 1990).

In order to face these problems, and to experiment with a more flexible methodology, approaches based on data abstraction techniques have been followed. An example of such an approach is found in the design of the Scratchpad II system (Jenks and Trager 1981).

This approach is a first step towards the design and development of an OOP system for the manipulation of mathematical data structures. In fact, the simple addition of data abstraction techniques is not sufficient. The usual features of systems must be enlarged in terms of logic capabilities, to satisfy another need of symbolic computation, that is the support of the manipulation of logic properties of data structures. These properties can be used both for the definition of the structures and for their direct use.

The selection of a suitable deductive tool and its integration into the defined

OOP environment are discussed in this paper, through examples of the tool's use as a programming language construct.

Regarding the existing automated deduction approaches, the connection method (CM) has been selected: its characteristics are stressed, in this paper, mainly comparing it with power and efficiency of the well known tableaux proof method (TPM); in this way, from the expressiveness point of view the equivalence between the two methods is proved (limiting the scope to propositional calculus in order to avoid unworthy cumbersome notation) and the main features of the connection method are shown.

A description of a prototypal implementation, related to the use of OOP methodology for the specification of mathematical data structures and for the treatment of first-order logic (FOL) properties is also given.

However this paper is not intended to provide a full description of the implementation aspects of such a module. The discussion will be focused on the programming methodology aspects, regarding the integration of deduction into symbolic computation and, as an intermediate means, into an OOP language.

2 The connection method

The first task of our work in this area has been to find a method for automated deduction in order both to fully integrate it with the system, to allow manipulation of logical properties, and to use it for theorem proving, in an interactive way, to deduce new properties of axiomatizable objects.

So the distinctive features we asked for an automated deduction method were based on the following considerations.

1. The structure of a given problem should be preserved during the development of a proof, in order to maintain a clean vision of the problem itself at each step of the deductive procedure. This helps the user, in not seeing great changes in its presentation of the problem ("human orientation").

2. The procedure should allow for further manipulation of a proof, towards development of proof presentation within the formalism of Gentzen's Natural Calculus; this characteristic can be very useful in using interactively the deductive system, giving another "human oriented" feature.

3. The method has to be efficient from the memory requirement point of view; note that this made acceptable its implementation through abstract data structures in a very high level programming language and makes reasonable the use of implemented module in the programming language.

We found these features in the so-called CM (Bibel 1987). The CM works in FOL and is based on the characterization of well-formed formulae (wff) as matrices that are sets of sets of . . . sets of atoms (literals in the ground case).

In the following this method is introduced through its close relationship with the TPM, showing the equivalence between these mechanisms from a constructive point of view, and the technical advantages of the former, with regard to space-efficiency and its adaptability to an object-oriented implementation. Ground formulae in disjunctive normal form (DNF) will be considered: this is

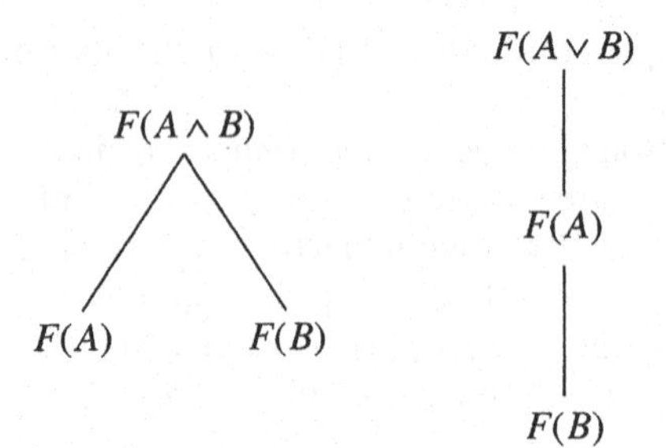

Fig. 1. Two rules of the tableaux proof method

for the sake of simplicity; this restriction, however, is without any loss of generality and our work has evolved towards a non clausal form; the extension to the predicative case is easy and natural, once the concepts of skolemization and unification have been introduced.

First a short description of the TPM is given, in order to introduce and prepare the next comparison with the connection method.

It should be noted that the following isn't the classical formulation of the TPM; it is a restriction due to the use of formulae in DNF; anyway it's well known that a formula D in DNF is valid if and only if it is provable with the TPM.

A proof delivered by the TPM can be described as the construction of a tree by means of two fundamental rules of Fig. 1. The next step is finding a set of pairs *SCP* (set of complementary pairs *CP*) of the form $\{F(A), F(\neg A)\}$ such that every branch of the tree contains at least one of them.

A generic formula in DNF can be represented by

$$D = C_1 \vee \ldots \vee C_n \tag{1}$$

where

$$C_i = L_1^i \wedge \ldots L_{k_i}^i \tag{2}$$

with k_i the number of literals in the i-th clause. Now the branches of a tree which has been constructed by the two rules of the TPM can be represented as in the following:

$$\beta = L_{j_1}^1, L_{j_2}^2, \ldots, L_{j_n}^n, \quad j_i \in [1, \ldots, k_i] \quad (i = 1, \ldots, n)\ . \tag{3}$$

So the following condition is called of *valid formula* (or *contradictory tree*):

$$\forall \beta, \exists CP \in SCP \quad \text{such that} \quad CP \subset \beta\ . \tag{4}$$

As an example let us consider the formula

$$(A \wedge \neg B) \vee (B \wedge C) \vee (\neg C \wedge B) \vee E \vee \neg A\ .$$

The validation tableaux and *SCP* are given in Fig. 2 with branches not completed because a complementary pair in them has already been detected.

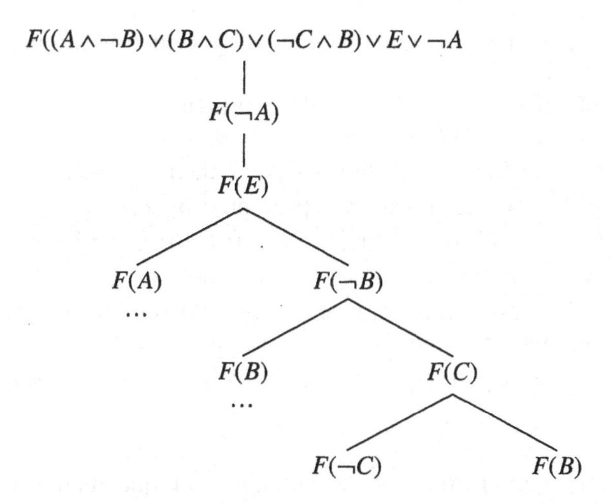

Fig. 2. $SCP = \{(\neg A, A), (\neg B, B), (C, \neg C), (\neg B, B)\}$. Dots represent branches not completed

Later we'll add some considerations about the role of literal E for the validation of this formula: actually it is useless in the proof, and we will see this as a direct result of proving by connections.

In the CM the operation of validation of a formula is performed by working on a matrix structure which represents the formula itself; here the definition of propositional matrix is given:

Definition 1. A (propositional) matrix F, on $\mathcal{P}^0$ (set of propositional symbols) and $\mathcal{R}$ (set of occurrences), together with its dimension $\delta(F)$, set of positions $\Omega(F) \subset \mathcal{R}$ and depth $\gamma(r)$ for every $r \in \Omega(F)$ are defined by induction through the laws:

1. for every literal L and for every $r \in \mathcal{R}$ the pair (L, r) is a matrix with $\delta((L, r)) = 0$, $\Omega((L, r)) = \{r\}$, $\gamma(r) = 0$;

2. if $F_1, \ldots, F_n$ with $n \geq 0$ are matrices such that $\Omega(F_i) \cap \Omega(F_j) = \{\}$ with $i \neq j$ and $1 \geq i, j \geq n$ then the set $\{F_1, \ldots, F_n\}$ is a matrix, say F, such that $\delta(F) = 0$ if $n = 0$ and $\delta(F) = 1 + \sum_{i=1}^{n} \delta(F_i)$ if $n > 0$, while $\Omega(\{F_1, \ldots, F_n\}) = \Omega(F_1) \cup \ldots \cup \Omega(F_n)$, and the depth of an occurrence r of a literal L in F is $\gamma(r) = m + 1$ if $r \in \Omega(F_i)$ and m is the depth of r as occurrence in F_i.

The atomic part of a matrix can be then either a simple literal or a nested set of occurrences of literals (in other words, an atom can be a literal or a nested matrix).

Then, matrices and formulae are related according to the next definition.

Definition 2. For every matrix F and for $l, m \in \{0.1\}$ the set of formulae $\tilde{F}$

represented by F respect to (l, m) is defined by induction through the following rules r1, r2, r3, r4:

r1. if F is a literal, $F = (L, r)$, and if $l = 0$ then $\tilde{F} = L$;

r2. if F is a literal, $F = (L, r)$, and if $l = 1$ then $\tilde{F} = \neg L$;

r3. if $F = \{F_1, \ldots, F_n\}$, $n > 0$ and if $m = 1$ then $\tilde{F} = (\tilde{F}_1 \wedge \ldots \wedge \tilde{F}_n)$ where the $\tilde{F}_i$, $i = 1 \ldots n$, are formulae represented by F_i w.r.t. $(l, 0)$;

r4. if $F = \{F_1, \ldots, F_n\}$, $n > 0$ and if $m = 0$ then $\tilde{F} = (\tilde{F}_1 \vee \ldots \vee \tilde{F}_n)$ where the $\tilde{F}_i$, $i = 1 \ldots n$, are formulae represented by F_i w.r.t. $(l, 1)$.

A formula $\tilde{F}$ is positively represented (by convention we say that it *is represented*) by the matrix F if it is represented by F with $l = m = 0$.

A formula $\tilde{F}$ is negatively represented by F if it is represented by F with $l = m = 1$.

In general, for every formula $\tilde{F}$ we obtain a unique matrix F such that $\tilde{F}$ is represented by F. On the other hand a matrix represents a set of logically equivalent formulae (being different one from another only with respect to the associativity and commutativity rules of the logical connectives $\vee$ and $\wedge$).

Among all matrices let us consider the particular set of the matrices in normal form, defined as follows.

Definition 3. A matrix F is said to be in *normal form* if $\max\{\delta(r)/r \in \Omega(F)\} \leq 2$. The elements of a matrix in normal form are called *clauses*.

Referring to the conventional way of representing a formula the set of matrices in normal form fully represents the set of formulae in DNF. It is now clear the choice of the name *matrix*: if every clause is a column of a bidimensional matrix, every literal of a clause becomes an element of a matrix in the sense of linear algebra, and this is the form we are interested in.

Another key concept for the CM is that of a path through a matrix:

Definition 4. A path π through a matrix F is a set of (occurrences of) literals defined by induction as follow:

p1. if $F = \{\}$ then the only path through F is the empty set $\{\}$;

p2. if $F = (L, r)$ then the only path through F is the set $\{(L, r)\}$;

p3. if $F = \{F_1, \ldots, F_m, F_{m+1}, \ldots, F_{m+n}\}$ with $0 \leq m, n,\ 1 \leq m+n,\ F_1, \ldots, F_m$ literals and $F_{m+1}, \ldots, F_{m+n}$ matrices that are not literals then the set

$$\bigcup_{j=1}^{m} \{F_j\} \cup \bigcup_{i=1}^{n} \pi_j$$

is a path through F if π is a path through E_i, for every matrix E_i belonging to F_{m+i}, with $1 \leq i \leq n$.

Let D be the formula as in (1) and (2); the matrix representation of D is

$$\{\{L_1^1, L_2^1, \ldots, L_{k_1}^1\}, \{L_1^2, L_2^2, \ldots, L_{k_2}^2\}, \ldots, \{L_1^n, L_2^n, \ldots, L_{k_n}^n\}\} .$$

According to the previous definitions, the form of a path through D is

$$\pi = \{L_{j_1}^1, L_{j_2}^2, \ldots, L_{j_n}^n\}, \quad j_i \in [1, \ldots, k_i] \quad (i = 1, \ldots, n) \tag{5}$$

the same as branches in the tableaux case (see (1)).

If we now define:

- a connection (C) is a pair of complementary literals;
- a set of connections (SC) is *spanning*, if in each path through the matrix there exists at least one connection belonging to that set;

then we can affirm that a formula D in DNF is valid if and only if a spanning set of connection (SSC) does exist for its matrix representation (saying, in this case, that the matrix is *complementary*):

$$\forall \pi, \exists C \in SSC \quad \text{such that} \quad C \subset \pi . \tag{6}$$

About the equivalence between TPM and CM, the following proposition can be stated:

Theorem 1 (Equivalence). Each formula D expressed in disjunctive normal form has a tableaux proof iff it has a representation in a matrix which is complementary.

The proof is constructively obtained by comparing the (4) and (6) that are the conditions for the successful validation according to the two methods respectively:

Proof. When D is fixed as in (1) and (2) it is easy to recognize that the characterizations of a branch β [(3)] and of a *path* π [(5)] are matching; the same holds for the definitions of complementary pair CP w.r.t. connection C and of set of complementary pair SCP w.r.t. spanning set of connections SSC.

Once the previous equivalences are seen, it is trivial to observe that conditions (4) and (6) state the same properties for two different representations.

□

Referring to the previous example, a validation of the formula of Fig. 2 can be given on the matrix of Fig. 3, where the spanning set of connections is

$$\{(\neg A, A), (\neg B, B), (C, \neg C), (\neg B, B)\} .$$

In this set each connection is between elements of two distinct clauses (columns) and useless clauses are simple to be detected by just listing the untouched columns. In our case the given formula is validated without using the

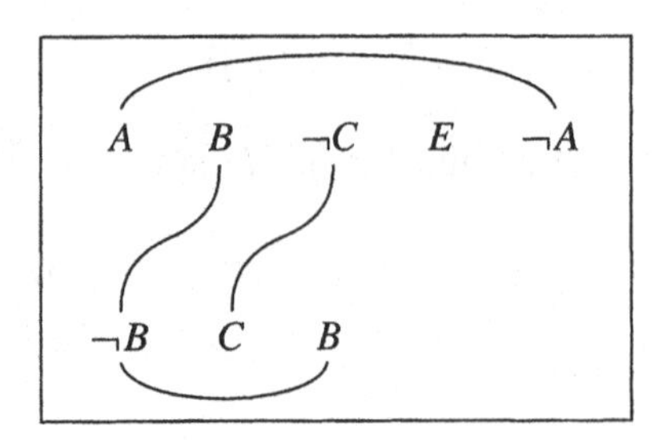

Fig. 3. Example of validation

E clause in any connection: if E is taken away the resulting formula can be validated as well.

Although the two methods have the same power, some differences have to be stressed, from an implementation point of view. The main difference between the TPM and the CM is that the matrix form for a formula gives many technical advantages, mainly avoiding redundancies, because the matrix remains untouched during the proof (and this advantage remains also with respect to the majority of resolution based methods).

The CM is simply extendible to full predicate logic, starting from its definition in the ground case, an appropriate integration of the previous methods and concepts. So the general matrix is the *quantified matrix*, on which the definition of path is extended; the introduction of the *unification concept* allows to extend analogously the definitions of connection, spanning set and complementary matrix to be used on first order atoms and quantified matrices. It is not significant, here, to give all these definitions, as the previous description of ground case was informally given, but it is worth to introduce the concept of multiplicity of a proof.

Definition 5. For a given F (formula not necessarily in DNF) a *multiplicity* μ is a function assigning to each node r in F (each root of a sub-formula of the form $\exists x_1 \ldots x_n\{F'\}$ for at least one F' in F) a natural number $\mu(r) \geq 1$.

As a matter of fact, to validate a formula it can be necessary to use more than one copy of the related matrix (just in a virtual way, however); the multiplicity μ of an attempted proof determines the number of copies of a sub-formula $\exists$ – *quantified* to be considered.

3 Implementation point of view

We have implemented a module for CM to be integrated in the programming system of project TASSO. It is still a prototype and it will give the requested logic capabilities to a basically imperative programming language. C++ has been used for implementation on a PC under Windows, i.e., in the same environment where a version of TASSO language is under development. An OOP approach to the implementation has been followed, sometimes having to deal with some kinds of incompatibilities of the Bibel algorithm with such methodology. Finally, the module is fairly an OOP system in which basic classes for matrix, clause,

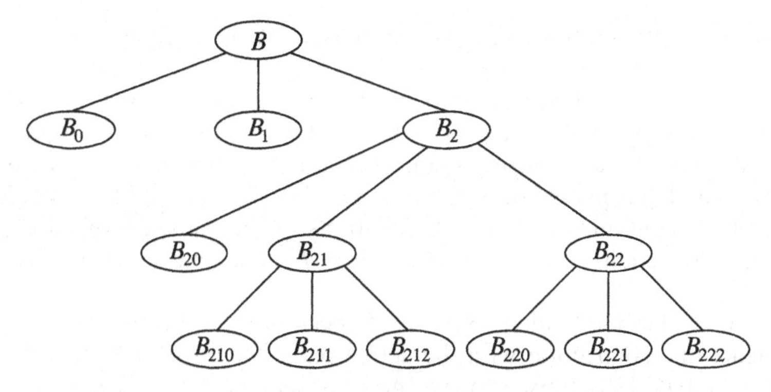

Fig. 4. Paths tree

literal, term, variable, constant are devised to model the main structures used by CM. In each class basic methods for construction, display, and destruction of objects and for proof delivery are provided.

Following Bibel (1987), the calculus proceeds through a sequence of steps each one being either an *extension*, a *separation*, or a *truncation*. We give a description of such procedure and display its effects on a general matrix.

Given such a general matrix, we indicate with B the set of all the possible paths on the matrix F, each made by k literals (if the matrix contains k clauses). During the application of the CM the tree of Fig. 4 is generated.

The initial extension step builds the root of the tree. A new partition of B, represented as a new level of the tree, is created, for each of the further extension steps, $B_{...} = \{B_{...0} \cup B_{...1} \cup B_{...2}\}$. The essence of this partition is described by the points 1 and 2, given below. The truncation steps consist in returning to the first non empty node $B_{...2}$ reachable going up in the tree. Separation steps have not to be considered because they produce a new demonstration in a reduced matrix.

With regard to Fig. 5, we call *active path* the set of all the paths of the matrix which have as common part the set of literals up to the current literal (the literal which is considered at the moment and which belongs to the current clause d). In Fig. 5, the d clause is the clause $c3$.

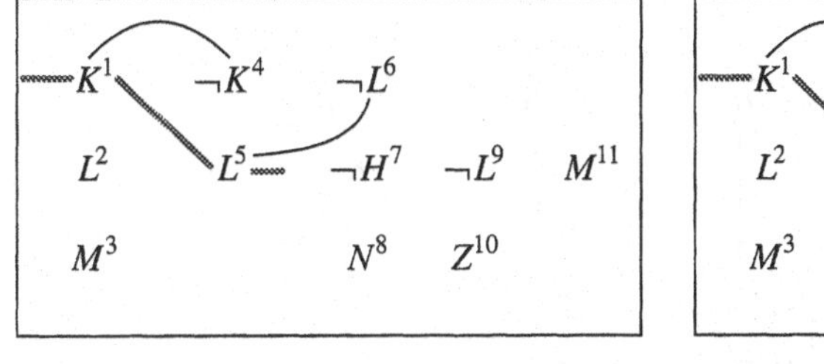

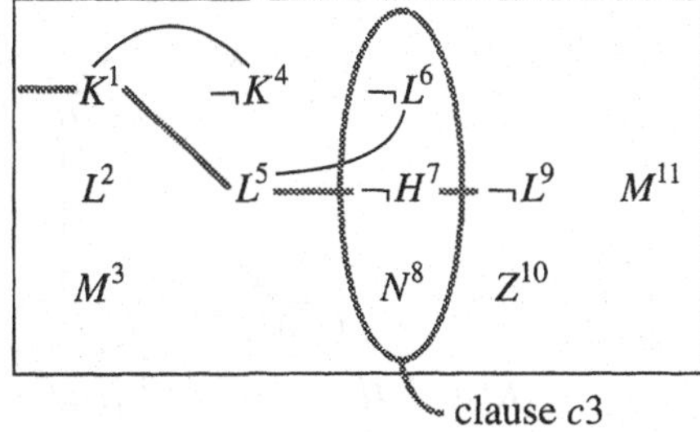

Fig. 5. Evolution of the active path (in grey)

The active path changes for every extension step as follows.

1. A new literal L_n, taken from current clause d, is added to the old active path p_a, so obtaining p'_a. In this way, two subsets of the set of paths are defined: those passing through L_n and those passing through the other literals of d, different from L_n. The latter paths are in the new set $B_{\dots 1}$. For example, in Fig. 5, if the active path is continued on $\neg H^7$, two subsets of the paths passing for $\{K^1 L^5\}$ are defined, depending on passing through $\neg H^7$. The paths passing through N^8 define $B_{\dots 1}$.

2. A new clause c (called $c3$ in Fig. 5) becomes the current one. This clause has at least one connection with the active path. Moreover, the first of the two sets of step 1 is decomposed in two new subsets. One of these, $B_{\dots 2}$, is made with all the paths passing through the literal L_n and a literal of c, having no connections with the active path. The other set is made by all the complementary paths and defines the new set $B_{\dots 0}$.

The partitions obtained through steps 1 and 2 by the matrix of Fig. 5 are shown in Fig. 6. The above process stops when at least one out of the following conditions is satisfied:

- a path with as many clauses as the matrix is built and, in this case, the matrix is non complementary;
- a tree, whose leaves, $B_{\dots 0}$, $B_{\dots 1}$, $B_{\dots 2}$, are empty, is obtained, and, in this case, the matrix is complementary.

The process terminates because the cardinality of the set of paths is finite and every extension step generates sets with less cardinality or, possibly, sets of the same cardinality which have not to be further partitioned.

Here we come to a crucial aspect of this implementation work, that is the management of the alternatives at each step.

First we consider that, once we have to choose among several literals for a further extension, then the neglected literal must be left to be eventually reconsidered. A WAIT stack is used for such purpose.

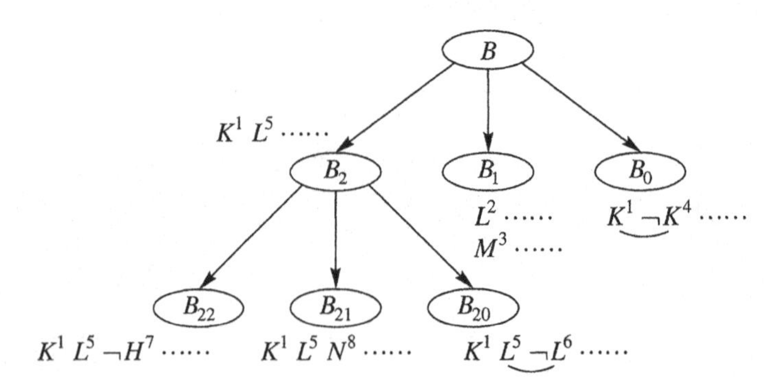

Fig. 6. Partitions of paths

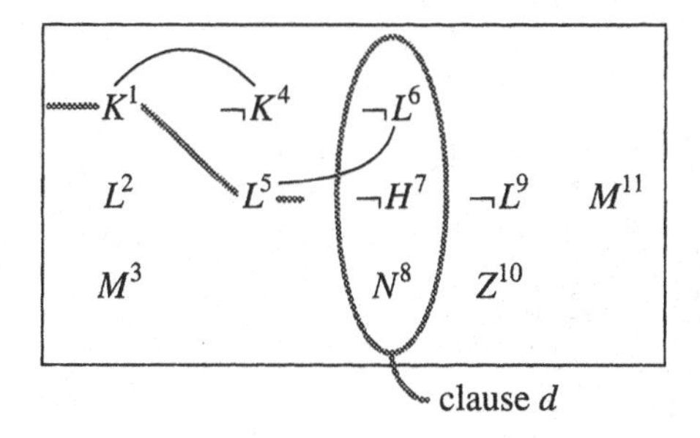

Fig. 7. Alternatives for further extensions

We can now add to the active path one out of the literals $\neg H^7$, N^8, as depicted in Fig. 7. The complementarity can be proved in a faster way, depending on this choice. On the other hand, in order to prove complementarity one must consider all the different alternatives. So, after the choice of the literal $\neg H^7$, the literal N^8 is still to be considered, unless a non complementary path has been found. This situation is called *alternative of type WAIT*.

Moreover, two other types of alternative can be considered. Note that they are really alternative to each other: once a decision has been taken we do not need to consider the unselected alternatives any more. Actually, to reach or to not reach an existing proof is not dependent on such alternatives. On the other hand, they may affect the speed of the proof. The *alternatives of first type* refer to the initial step of extension, when the clause, from which the proof starts, has to be chosen. The *alternatives of second type* refer to the choice of a clause for the extension, that must contain at least one connection with a literal in the active path.

In Fig. 8 the active path p_a, with the literals K, L and M, is shown. Extensions can be performed towards clauses $c4$, $c5$, $c6$. Only $c4$ and $c5$ can be chosen, in fact, because $c6$ has no connection with the active path.

The problem of the alternatives can be well represented by a tree, whose arcs are labeled by the type of relative alternatives.

The tree of Fig. 9 shows the *tree of the total alternatives* for the matrix of Fig. 5. In the propositional case, each one of the partial trees contains a terminal node, labeled by *"end"*.

In order to deal with predicative formulae, *alternatives of third type* have to be considered. After the clause for the extension has been selected, the unification problem has to be faced.

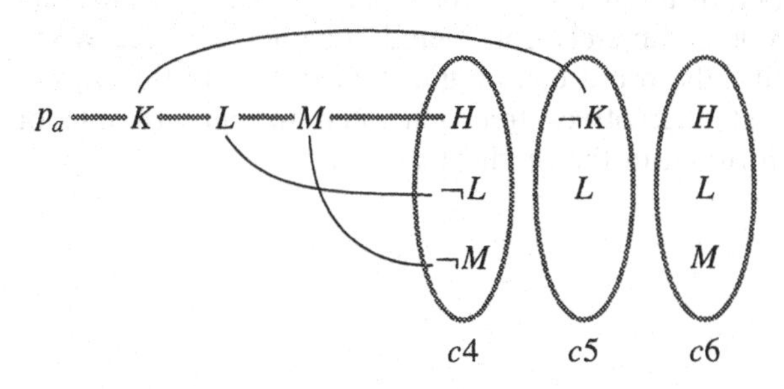

Fig. 8. Alternatives of second type

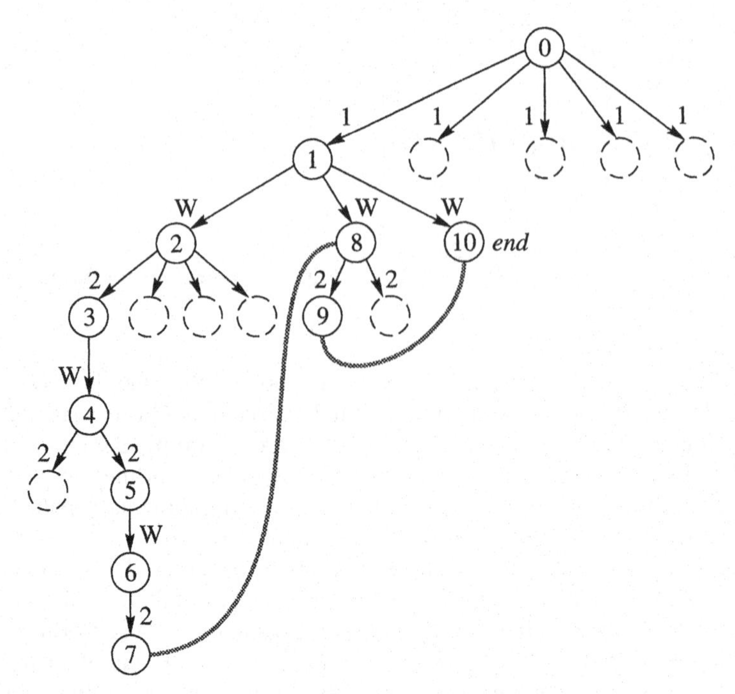

Fig. 9. Total alternatives tree for propositional case. Node number states order of tree visit. Broken, nodes not considered during the process. In gray, truncation steps

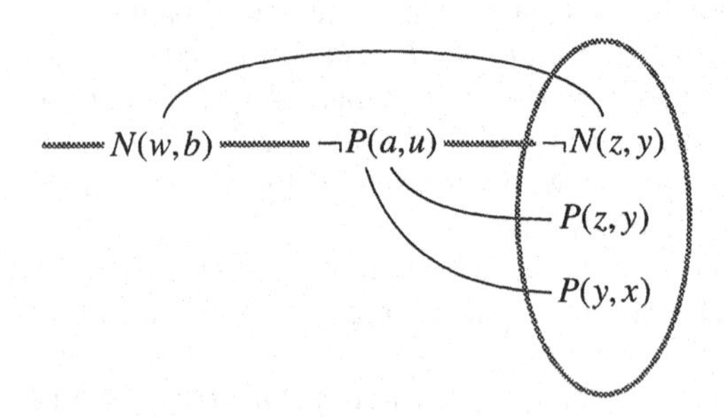

Fig. 10. Active path in predicative case. Solidly arced, clauses chosen for the extension. In gray, active path

In Fig. 10, since we have three possible connections, six different unifiers have to be considered. If we extend the use of the alternatives tree also to the predicative case, this important difference has to be dealt with. All the different types of alternatives have to be completely considered, together with the WAIT alternatives. Actually, pursuing the proof can be guaranteed only in this way.

Figure 11 shows the total alternatives tree generated for an example of demonstration of a generic formula in the predicative case.

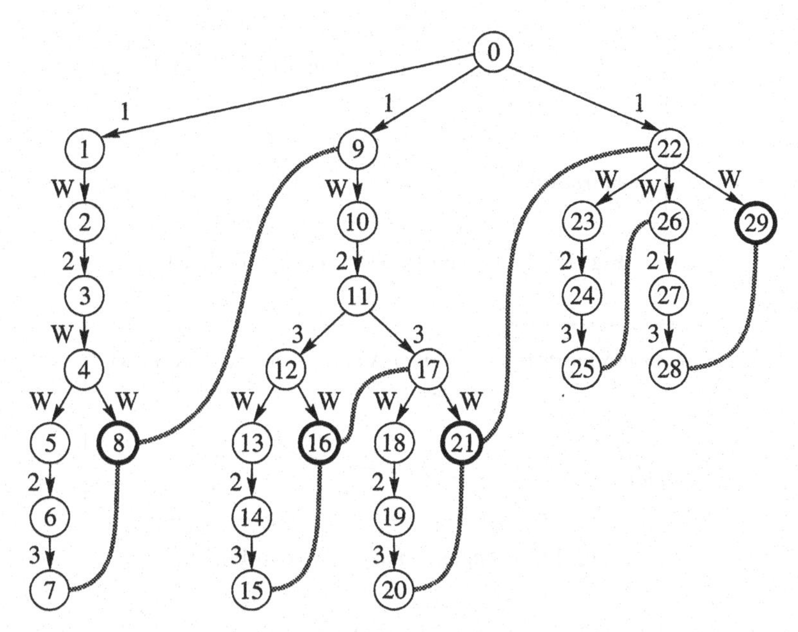

Fig. 11. Example of total alternatives tree for predicative case. Node number states order of tree visit. In bold, nodes where creating a copy of the original matrix is necessary. In gray, truncation steps

4 Uses of the deductive system

Uses of the module, embedded into programming system TASSO, are thought for reasoning about classes defined through the language at a static stage, and for evaluation of properties of objects during program execution. In TASSO language, software for symbolic computation can be designed and implemented by OO language constructs; then, the logic constructs, allowed by our deductive tool, enable to look for properties of given objects before trying some given computation.

Even if still apart from the programming language, the module is in use and has been tested over a variety of problems in mathematical data structures manipulation. Formulae from Pelletier (1986) have been exploited for testing purposes.

Moreover, the programming techniques on which this implementation is based allow for more interesting developments, by means of few modifications. Here we discuss a couple of possible extensions that we consider of particular interest for the development of the tool.

An interesting application concerns the possible completion of a set of hypotheses: starting from a given formula, represented by a non complementary matrix, a new complementary matrix could be obtained just by adding some selected atoms to the old one. The new atoms are added in the paths that were without connections in the old matrix, such that each one of them will contain at least one connection in the new matrix. As an example the well known inference problem of the mortality of Socrates can be mentioned: we can start from a not

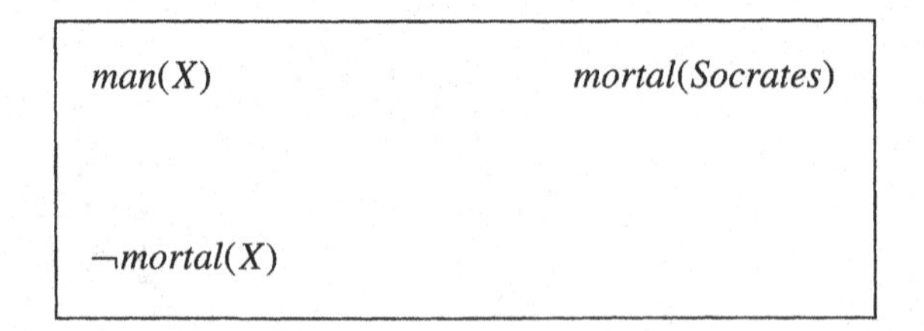

Fig. 12. Example of matrix to be completed

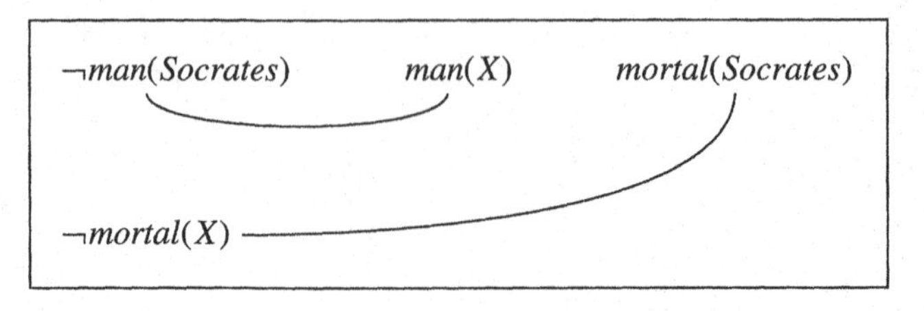

Fig. 13. Complementary matrix to Fig. 12

complete formulation, by a set of one single hypothesis

$$\forall X\colon\ man(X) \Rightarrow mortal(X)$$

and the well known thesis *mortal(Socrates)*. We can try to prove the thesis starting from that single hypothesis; this means to see whether

$$(\forall X\colon\ man(X) \Rightarrow mortal(X)) \Rightarrow mortal(Socrates)\ ,$$

that is to verify the simply reachable formula

$$(man(X) \wedge \neg mortal(X)) \vee mortal(Socrates)$$

which has the matrix representation given in Fig. 12.

An attempt to reach a spanning set of connections results unsuccessfully; this depends on the lack of an atom to be connected (and unified) with $man(X/Socrates)$. The result of such a remark can be the addition of a new atom to the matrix, of the form $\neg man(Socrates)$ that is a supplementary hypothesis to $man(Socrates)$; the resulting matrix (Fig. 13) is complementary and represents a minimal extension of the original formula to the following valid one

$$\neg man(Socrates) \vee (man(X) \wedge \neg mortal(X)) \vee mortal(Socrates)$$

whose sense is

$$(man(Socrates) \wedge (\forall X\colon\ man(X) \Rightarrow$$
$$\Rightarrow mortal(X))) \Rightarrow mortal(Socrates)\ .$$

In other words, we establish that, once the hypothesis *man(Socrates)* can be

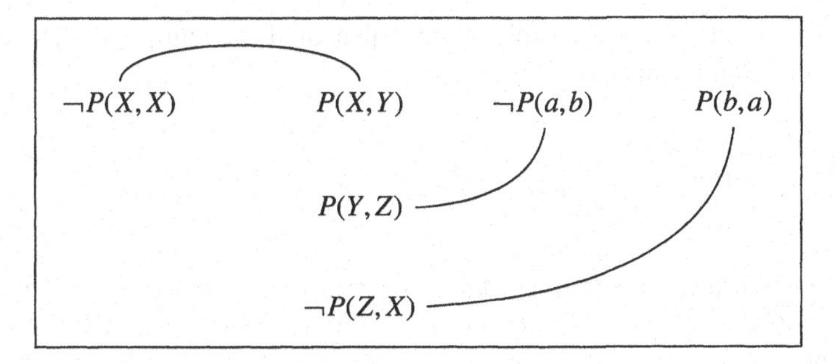

Fig. 14. Complementary matrix

added without corrupting the soundness of the whole set of hypotheses, then a valid formula is obtained. This type of application is meaningful towards applications concerning the use of *abduction* in automated reasoning (Eshghi and Kowalski 1989).

A second feature worth to be mentioned is the fact that the implemented module can be already used for verifying mathematical properties of objects treated by a programming language.

Let us suppose to have a graph to be manipulated, and that such manipulation needs the graph, say, to be symmetric. If the following properties are given among the nodes of the graph

$$\forall X\colon\ P(X, X)\ ,$$
$$\forall X, Y, Z\colon\ P(X, Y) \wedge P(Y, Z) \Rightarrow P(Z, X)$$

the symmetry property $\forall a, b\colon\ P(a, b) \Rightarrow P(b, a)$ can be verified by validating the resulting formula

$$\neg P(X, X) \vee (P(X, Y) \wedge P(Y, Z) \wedge \neg P(Z, X)) \vee \neg P(a, b) \vee P(b, a)$$

represented by the complementary matrix given in Fig. 14.

A more clean use can be reached for the deductive module, with respect to the last example, once it is embedded into the programming system. The module can become a system service designed to work on instances of variables of type *formula*: there is sense in attempting to embed such a module into the resulting programming language of the project and an example of such a use can be shown as follows.

If a *Graph* definition is given, with attributes representing properties evaluable on instances, and the variables *symmetry* and G are defined as

VARIABLE *symmetry*: formula;
VARIABLE G: *Graph*;

and assigned as *symmetry* := $[\forall a, b\ P(a, b) \Rightarrow P(b, a)]$ and G := new *Graph*; with further assignments on attributes of the object G, in order to state the hypotheses given in the example, then a use of the eventual system operation `proof` can be presented as

if `proof` $(G, symmetry)$
then (manipulation of G)
fi

The possibilities offered by the programming system and methodologies allow for next extensions by an incremental logic. Two main directions for such extensions are:

- the possibility to perform parallel computations (namely by creating a new process each time the multiplicity is incremented).
- the improvement of the unification mechanism, that could be implemented by a more efficient algorithm (Paterson and Wegman 1978).

With regard to the efficiency of the general algorithm for the connection method, we expect to develop some enhancements, such as adding the *splitting by need* mechanism and supporting induction and use of equality (Bibel 1987).

Acknowledgement

This work has been partially supported by CNR under the project "Sistemi Informatici e Calcolo Parallelo", under grant no. 92.01604.69.

References

Bibel, W. (1987): Automated theorem proving. F. Vieweg und Sohn, Wiesbaden.
Eshghi, K., Kowalski, R. (1989): Abduction compared with negation by failure. In: Levi, G., Martelli, A. (eds.): Proceedings Sixth International Conference on Logic Programming, ICLP. MIT Press, Cambridge, MA, pp. 234–254.
Fitch, J. (1985): Solving algebraic problems with Reduce. J. Symb. Comput. 1: 211–227.
Jenks, R. D., Trager, B. (1981): A language for computational algebra. SIGPLAN Not. 16/11: 22–29.
Miola, A. (ed.) (1990): Design and implementation of symbolic computation systems. Springer, Berlin Heidelberg New York Tokyo (Lecture notes in computer science, vol. 429).
Paterson, M., Wegman, M. (1978): Linear unification. J. Comput. Syst. Sci. 16: 158–167.
Pelletier, F. (1986): Seventy-five problems for testing automatic theorem provers. J. Autom. Reason. 1: 191–216.
Wang, P., Pavelle, R. (1985): MACSYMA from F to G. J. Symb. Comput. 1: 69–100.

A general reasoning apparatus for intelligent tutoring systems in mathematics

A. Colagrossi and A. Micarelli

1 Introduction

Since the early sixties the use of computers in education (computer aided instruction, CAI) has aimed to individualize instruction, i.e., tailor study procedures to the needs and characteristics of the individual student. However, although some progress has been made, CAI is still a long way off from achieving this goal. In practice, with CAI systems the student has a passive role, being exposed to a fixed set of problems that have been stored together with a corresponding set of pre-canned solutions. The tutoring path is rigid, with one-way teaching interaction only and with very little individualization. The possibility to get over the limits of the traditional CAI came from studies on cognitive psychology and artificial intelligence, with particular attention paid to the interaction between teacher and student. These researches led to the development of new educational systems, called intelligent computer aided instruction (ICAI) systems or intelligent tutoring systems (ITS), where artificial intelligence techniques are largely applied.

Intelligent tutoring systems are expert systems in education. Traditionally, four main modules have been identified in which knowledge is explicitly represented in ITS: the *expert module*, capable of actually solving problems in the domain to be taught; the *tutoring module*, which embeds the teaching expertise (teaching strategies and diagnostic capabilities); the *student modeling module*, which builds an explicit representation of the learning status of the student, and the *user interface* (Sleeman and Brown 1982, Wenger 1987, Yazdani 1986). One of the main features of ITS, compared to traditional CAI systems, is their ability (embedded in the expert module) to simulate expert problem-solving in the domain to be taught. This feature has some important consequences in the performance of ITS. In particular, if the tutoring system is endowed with problem solving capabilities, the student may have an active role, being allowed to take the initiative by not only asking for explanations but also by proposing new problems thus realizing the so-called *mixed initiative* dialogue. At the state of the art, no general architecture for ITS has been identified as yet. Nevertheless, there are design principles that are common to all the architecture of ITS. They are relative to the method of teaching (which essentially consists in breaking the teaching action down into diagnosis, repair planning and step-by-step monitoring) and to the fact that the system is knowledge based: it embodies an explicit representation of the knowledge about the student, the tutoring strategies, the curriculum and, above all, the knowledge about the domain to be taught. As far as ITS for teaching mathematics are concerned, several systems have been developed over the years, probably more than in any other domain. In fact, mathematics is a highly structured domain that constitutes the background for most science and engineering, a valid test domain and, as a consequence, an important source of inspiration for ITS research. In this paper, we propose a general reasoning module for supporting the various functionalities of ITS, such as autonomous problem solving on the domain to be taught, diagnostic capabilities, monitoring, etc. For example, it can be used for the diagnostic functionality which is triggered when wrong solutions are proposed by the student. Its task is to pinpoint student mistakes present in her solution and find causes for them. The reasoning apparatus we have chosen is an extended sequent calculus, particularly suitable for educational purposes. In fact, this inference tool is natural (i.e., proofs are in a style close to the human reasoning), a feature that facilitates the step-by-step monitoring of the solution procedure of a student. Another interesting feature is its uniformity in the solution of different kinds of problems involving the manipulation of mathematical objects, problems we consider pedagogically prominent due to the various skills they need in students. In fact, an ITS for mathematics is intended to stimulate the student to manipulate properties of mathematical objects rather than only compute with them. The selection of mathematical problems to be proposed to the student must be driven by the objective of enhancing his reasoning attitude rather than only improving his technical and mechanical computing skill. Mathematical problems, independently of the particular domain they belong to, can be classified as *verificative* problems, requiring the verification that a property holds for objects in a given theory, *generative* problems, requiring the generation of a property as consequence of the fact that other properties hold for objects in a given theory, and *abductive* problems, requiring the abduction of the hypothesis from which a given property holds. In the paper we present a description of ITS in mathematics, a description of the sequent calculus and show an example of the use of the reasoning apparatus in ITS, stressing the advantages of the proposed approach for the construction of ITS.

2 ITS in mathematics

Table 1 shows some of the most known ITS in mathematics and summarizes their prominent characteristics. Apparently they use different formalisms for knowledge representation and different approaches in the teaching/learning relationship. We look at two of them in some details. LMS/Pixie developed by Sleeman (1983, 1987) is a system for teaching students how to solve linear algebraic equations (e.g., $6 * x + 5 = 8 - 3 * x + 12$). The knowledge on the domain to be taught has been represented by means of production rules, used in forward chaining. Table 2 lists the rules used in LMS/Pixie (*lhs* and *rhs* are gen-

Table 1. Characteristics of intelligent tutoring systems

System	Subject matter	Language	Knowledge base	Reference
Buggy	subtraction	Lisp	procedural network	Brown and Burton 1978
West	arithmetic expressions	PLATO system	rule based representation	Burton and Brown 1979
Quadratic	quadratic equations	Lisp	self-improving rule based representation	O'Shea 1982
LMS/Pixie	algebraic procedures	Lisp	rule and mal-rule based representation	Sleeman 1983, 1987
Geometry Tutor	rational geometry	OPS5	goal-restricted production rules	Anderson et al. 1985
Amalia	algebraic calculus	Lisp	rule based representation	Vivet 1987
Sedaf	study of math functions	KEE and Lisp	rule, mal-rule, and frame based representation	Aiello and Micarelli 1990

Table 2. Rules in LMS/Pixie

Name	Level	Condition/action
Fin2	1	$(x = M/N) \Longrightarrow ([evaluated])$.
Solve	2	$(M * x = N) \Longrightarrow (x = N/M)$ or $(INFINITE)$.
Addsub	3	$(lhs\ M +/- N\ rhs) \Longrightarrow (lhs\ [evaluated]\ rhs)$.
Mult	4	$(lhs\ M * N\ rhs) \Longrightarrow (lhs\ [evaluated]\ rhs)$.
Xaddsub	5	$(lhs\ M +/- N\ rhs) \Longrightarrow (lhs\ [evaluated]\ rhs)$.
Ntorhs	6	$(lhs +/- M = rhs) \Longrightarrow (lhs = rhs -/+ M)$.
Rearrange	7	$(lhs +/- M +/- N * x\ rhs) \Longrightarrow$ $(lhs +/- N * x +/- M\ rhs)$.
Xtolhs	8	$(lhs = +/- M * x\ rhs) \Longrightarrow (lhs -/+ M * x = rhs)$.
Bra1	9	$(lhs\ \langle N \rangle\ rhs) \Longrightarrow (lhs\ N\ rhs)$.
Bra2	10	$(lhs\ M * \langle N * x +/- P \rangle\ rhs) \Longrightarrow$ $(lhs\ M * N * x +/- M * P\ rhs)$.

eral patterns, and M and N are integers). They can be viewed as rewrite rules. For instance, the first one (Xtolhs) moves x terms on the right-hand side of an equation into the left-hand side of the equation, changing the sign in the process. The conflict resolution strategy of the above production system takes into account only the priority of the application of rewriting rules (e.g., multiplication

Table 3. Mal-rules in LMS/Pixie

Name	Level	Condition/action
Msolve	2	$(M * x = N) \Longrightarrow (x = M/N)$.
Mntorhs	6	$(lhs +/- M = rhs) \Longrightarrow (lhs = rhs +/- M)$.
M2ntorhs	6	$(lhs1 +/- M\ lhs2 = rhs) \Longrightarrow$ $(lhs1 +/- lhs2 = rhs +/- M)$.
Mxtolhs	8	$(lhs = +/- M * x\ rhs) \Longrightarrow (lhs +/- M * x = rhs)$.
M1bra2	10	$(lhs\ M * \langle N * x +/- P\rangle\ rhs) \Longrightarrow$ $(lhs\ M * N * x +/- P\ rhs)$.
M2bra2	10	$(lhs\ M * \langle N * x +/- P\rangle\ rhs) \Longrightarrow$ $(lhs\ M * N * x +/- M +/- P\ rhs)$.

has to be done before addition). The common bugs are represented explicitly as variants of the correct rules. These variants are called mal-rules and are used by the system to model the student in the case of wrong answers (Table 3).

LMS/Pixie is a diagnostic modeling system: it infers from the student's solution which rules and mal-rules compose his/her solution, by also specifying the order of the application of the rules. During an interactive session, the student is presented with a sequence of "task sets" of growing complexity. For each problem, the system compares the student's answer with the solution, determined by the expert module, constituted by an ordered set of correct rules. If the proposed solution is correct, the system considers this ordered set of rules as the student model. If the solution is incorrect, the system identifies the model (ordered set of rules) which leads to the proposed solution, chosen among a set of models (also containing mal-rules, if that is the case) created by the system in an off-line phase, and considers this model as the student model. At this point, the remedial subsystem is activated: it gives a suitable remediation on the basis of the content of the student model, by personalyzing the teaching dialogue to the needs of the student.

The Geometry Tutor has been developed by Anderson et al. (1985). It is a system for teaching proof generation in elementary geometry. The knowledge base on the subject domain and the bug catalogue are represented in the form of goal-restricted production rules (the conditions of the rules include a specific goal). The expert module of the system contains 300 production rules, approximately. Two examples of rules follow:

if the goal is to prove $\Delta XYZ \cong \Delta UYW$
and X, Y, W are collinear
and U, Y, Z are collinear
then conclude $\angle XYZ \cong \angle UYW$ because of vertical angles.

if the goal is to prove $\Delta XYZ \cong \Delta UVW$
and $XY \cong UV$
and $YZ \cong VW$
then set a subgoal to prove $\angle XYZ \cong \angle UVW$ so as SAS can be used.

The same formalism is used to represent the common bugs in the subject domain. The didactic approach used by Geometry Tutor is cooperative: at the beginning of a session the student is presented with a geometric problem, graphically represented on the screen. During the session, the inferences made by the student are monitored step by step and graphically represented to indicate where the student is. If the student makes an error, the system prompts an immediate hint to guide him back to the correct path.

It can be stressed how all the mentioned systems are based on artificial intelligence technology. However, no clear general architecture for ITS has been identified as yet. All ITS are geared around an expert module that plays a crucial role for the didactic effectiveness of ITS. The development of such modules is often time consuming, since they are ad hoc for each specific class of subject domains. Thus, the effectiveness of such ITS is often limited by the limited size of the domain.

As for the functionality of the expert module (problem solving in the domain), one may think of using existing general purpose systems for mathematical problem solving, such as Macsyma (Moses 1975), Maple (Char et al. 1991) or Mathematica (Wolfram 1991). The domains they deal with are wide and they are very effective in solving mathematical problems by using methods like expansion, collection or application of formulae. However, they have not been designed to provide an explicit representation of the process that leads to the solution of a given problem (representation required for teaching strategic skills in the domain), so that it is not easy to have them reason about solution processes. More recent trends in the design of symbolic manipulation systems consist in endowing them with an explicit representation of mathematical knowledge to be used for reasoning, along with procedures to perform symbolic computations (Limongelli et al. 1991). According to this approach, a general architecture for ITS in mathematics can be envisaged as an enhancement of the reasoning capabilities of a symbolic computation system. In the rest of the paper we briefly present this architecture, where the reasoning apparatus is a sequent calculus, and show some examples.

3 The reasoning apparatus

In this section we will present a brief description of the sequent calculus which is the basis of the proposed reasoning apparatus. Here we describe only the features prominent for the use of the apparatus in ITS to make the examples presented in the following section more understandable. A whole description of the sequent calculus and the implementation of a software tool based on it can be found in Bertoli et al. 1997; Cioni et al. 1992, 1995, 1997). An application of this sequent calculus to ITS can also be found in Aiello et al. (1993). A sequent is a pair (Γ, Δ) of sets of well-formed formulae of first-order predicative logic. Γ is called antecedent, Δ is called succedent and the usual representation of a sequent is:

$$\Gamma \rightarrow \Delta .$$

The semantics of sequents can be expressed as "the conjunction of the formulae of the antecedent implies the disjunction of the formulae of the succedent". In symbols,

$$\gamma_1 \wedge \gamma_2 \wedge \ldots \wedge \gamma_n \supset \delta_1 \vee \delta_2 \vee \ldots \vee \delta_n \,. \tag{1}$$

Thus, a sequent is valid (or is a contradiction, or is satisfiable) if and only if formula (1) is valid (or is a contradiction, or is satisfiable). A sequent can be manipulated by inference rules. Here we present a short list of such rules:

$\underline{\neg\mathbf{A}}$: $\dfrac{\Gamma_1,\neg\alpha,\Gamma_2\rightarrow\Delta}{\Gamma_1,\Gamma_2\rightarrow\alpha,\Delta}$,

$\underline{\wedge\mathbf{A}}$: $\dfrac{\Gamma_1,\alpha\wedge\beta,\Gamma_2\rightarrow\Delta}{\Gamma_1,\alpha,\beta,\Gamma_2\rightarrow\Delta}$,

$\underline{\vee\mathbf{A}}$: $\dfrac{\Gamma_1,\alpha\vee\beta,\Gamma_2\rightarrow\Delta}{\Gamma_1,\alpha,\Gamma_2\rightarrow\Delta \quad \Gamma_1,\beta,\Gamma_2\rightarrow\Delta}$,

$\underline{\supset\mathbf{A}}$: $\dfrac{\Gamma_1,\alpha\supset\beta,\Gamma_2\rightarrow\Delta}{\Gamma_1,\Gamma_2\rightarrow\Delta,\alpha \quad \Gamma_1,\beta,\Gamma_2\rightarrow\Delta}$.

In the following two rules, $t, t_1, \ldots, t_n$ are terms of the theory of interest:

$\underline{\forall\mathbf{A}}$: $\dfrac{\Gamma_1,\forall x\alpha(x),\Gamma_2\rightarrow\Delta}{\Gamma_1,\forall x\,\alpha(x),\alpha(t_1/x),\ldots,\alpha(t_n/x),\,\Gamma_2\rightarrow\Delta}$,

$\underline{\exists\mathbf{A}}$: $\dfrac{\Gamma_1,\exists x\alpha(x),\Gamma_2\rightarrow\Delta}{\Gamma_1,\alpha(t/x),\rightarrow\Delta}$.

Two important characteristics hold for the inference rules. The first one is that they can be applied to any first-order well-formed formula, that is they don't require a process in order to put a given formula in a specific normal form. The second aspect is related to the naturalness of the inference rules. A sequent is decomposed into one or two sequents following a natural style, that is simulating the elementary logical steps of a reasoning process performed by a human being.

By applying the inference rules to a given sequent it is possible to determine its validity. In fact the following theorem holds: "a sequent is valid if the application of inference rules decomposes it into all basic valid sequents." We recall that a basic valid sequent is a sequent for which one of the following three cases holds:

1. one formula in the antecedent is a contradiction;
2. one formula in the succedent is a tautology;
3. there exists a pair of equivalent formulae, the first in the antecedent and the second in the succedent.

Cioni et al. (1997) present a reasoning procedure for a sequent. Such a procedure (procedure $\mathcal{R}$, in the following) is composed of two parts: the first is the proving part and allows for checking the validity of a given sequent. The second is the real reasoning part and it is activated only if the given sequent has been proved to be not valid. The output of the reasoning part is a pair of sets of formulae such that:

1. the sequent obtained by adding to the antecedent whatever formula of the first set (also called the abduction set) is valid;
2. the sequent obtained by adding to the succedent whatever formula of the second set (also called the generation set) is valid.

Such a reasoning procedure can be used for solving verificative, generative, and abductive problems.

A mathematical problem can be modeled considering:

1. a mathematical environment E, i.e., the set of theorems prominent for the considered problem;
2. a hypothesis A, i.e., a set of known properties that hold in the considered mathematical theory;
3. a thesis B, i.e., a set of properties that should be derived from A in the theory.

Let us consider the following sequent π:

$$E \rightarrow A \supset B \; .$$

If we consider E as the environment, A as the hypothesis, and B as the thesis of a problem P in a given theory T, then the sequent π models the problem P, in the sense that the sequent π is valid if B is a logical consequence of A, given E. Then, working with the sequent π, some information about the problem P can be obtained.

Using the methods presented by Cioni et al. (1997) we are able to solve three kinds of problems, called *verificative*, *generative*, and *abductive*, which are pedagogically prominent for teaching mathematics, due to the various skills they require in students.

Thus, the three kinds of problems can be formulated as follows.

1. A verificative problem is solved by applying the proving part of procedure $\mathcal{R}$. If $A \supset B$ is a logical consequence of E, then the proving terminates and the output is an empty set. Otherwise, either the procedure terminates with a non-empty set of formulae or the procedure doesn't terminate, due to the semidecidability of the first-order logic calculus.

2. An abductive problem is solved by applying the reasoning part of procedure $\mathcal{R}$ to the set of formulae output of the proving part. The output of the reasoning part is a pair of sets of formulae. Let α be a formula of the first set. The following sequent is then valid:

$$E, \alpha \rightarrow A \supset B \; .$$

3. A generative problem is considered as the dual of the abductive problem; in fact, it is solved in an analogous way. Let β be a formula of the second set of

the pair output of the reasoning procedure. The following sequent is then valid:

$$E \to \beta,\ A \supset B\ .$$

In the following we will use an enhanced version of the sequent calculus which allows to perform reasoning in a given domain (Bonamico et al. 1993). In particular, the rules about quantifiers have been modified in order to treat mathematical objects belonging to distinct domains.

4 An example

In this section we show, through a simple example, the use of the proposed reasoning apparatus for performing the various functionalities required in an ITS. Suppose that the student, during a teaching session, is asked to reason about the continuity property of the sum of two given mathematical functions, continuous in closed real intervals. In the following, the functions and the intervals are shown:

$$u\colon\ x^2+1 \qquad I_u\colon\ [-3,+3]$$

$$v\colon\ 2x^5-3x^2+2 \qquad I_v\colon\ [-1,+5]$$

Suppose T consists in the theory of real functions of a single real variable defined over closed intervals. The environment E consists in the only theorem on the continuity of the sum of two continuous functions:

For any pair of functions (f, g) and any pair of closed intervals (I_f, I_g), there exists a function h and an interval I_h such that if (f is continuous in I_f) and (g is continuous in I_g) and (h is the sum of f and g) and (I_h is the intersection of I_f and I_g) then (h is continuous in I_h).

The above theorem is easily expressed in first-order logic as follows:

$$\begin{gathered}(\forall f, g, I_f, I_g\ \ \exists h, I_h)\\ ((\mathrm{Cont}(f, I_f) \wedge \mathrm{Cont}(g, I_g) \wedge \mathrm{Int}(I_f, I_g, I_h) \wedge \mathrm{Sum}(f, g, h))\\ \supset \mathrm{Cont}(h, I_h))\end{gathered} \tag{2}$$

As was already stated, the reasoning apparatus supports the tutoring system in performing the following actions:

1. finding and showing the proof of a given problem;
2. monitoring the student's problem solving behavior;
3. detecting and recovering the student's mistakes;
4. suggesting the next step to be performed;
5. explaining the meaning of the steps performed.

Regarding the first point, finding and showing the proof, the expert module builds the following sequent:

$$(\forall f, g, I_f, I_g \ \exists h, I_h)$$
$$((\mathrm{Cont}(f, I_f) \wedge \mathrm{Cont}(g, I_g) \wedge \mathrm{Int}(I_f, I_g, I_h) \wedge \mathrm{Sum}(f, g, h)) \tag{3}$$
$$\supset \mathrm{Cont}(h, I_h)), \mathbf{Ob} \rightarrow$$

where **Ob** is a given observation, i.e., the facts that are assumed to hold in the considered theory. Afterwards, the system is able to perform the following steps:

Step 1. The $\underline{\forall \mathbf{A}}$ rule, relative to f, g, I_f, I_g, furnishes the following sequent:

$$(\exists h, I_h)$$
$$((\mathrm{Cont}(u, I_u) \wedge \mathrm{Cont}(v, I_v) \wedge \mathrm{Int}(I_u, I_v, I_h) \wedge \mathrm{Sum}(h, v, h))$$
$$\supset \mathrm{Cont}(h, I_h)), \mathbf{Ob} \rightarrow$$

Note that one of the features of the reasoning apparatus operated several simplifications applying the following criteria:

1. discard the formulae where particular cases occurred (e.g., $\mathrm{Int}(I_u, I_u, I_h)$ and/or $\mathrm{Sum}(v, v, h)$);
2. discard the formulae where incoherent values occurred (e.g., $\mathrm{Cont}(u, I_v)$ and/or $\mathrm{Cont}(v, I_u)$);
3. take into consideration only one representative formula among a class of formulae that are equivalent under the commutative property (in this case, the commutativity of the sum Sum and/or the intersection Int).

Step 1 corresponds to an instantiation of formula (2) with the given functions u and v and their respective domains.

Step 2. The system applies the $\underline{\exists \mathrm{A}}$ rule to the sequent derived in Step 1. Such a rule performs the computation of the sum of the functions u and v and the intersection of the intervals I_u and I_v. The following sequent is then derived:

$$((\mathrm{Cont}(u, I_u) \wedge \mathrm{Cont}(v, I_v) \wedge \mathrm{Int}(I_u, I_v, I_h) \wedge \mathrm{Sum}(h, v, h))$$
$$\supset \mathrm{Cont}(z, I_z)), \mathbf{Ob} \rightarrow$$

where z: $2x^5 - 2x^2 + 3$ and I_z: $[-1, +3]$.

Step 2 corresponds to the identification of the sum of the two given functions and of its domain.

Step 3. The application of the $\underline{\supset \mathbf{A}}$ rule generates the following sequents:

1. $\mathrm{Cont}(z, I_z), \mathbf{Ob} \rightarrow$;

2. **Ob** $\rightarrow$ $(\text{Cont}(u, I_u) \wedge \text{Cont}(v, I_v) \wedge \text{Int}(I_u, I_v, I_h) \wedge \text{Sum}(h, v, h))$.

At this point different tracks can be followed, depending on the data included in **Ob**. We suppose two different situations:

1. **Ob** is the following conjunction:

$$\text{Cont}(u, I_u) \wedge \text{Cont}(v, I_v) \wedge \text{Int}(I_u, I_v, I_h) \wedge \text{Sum}(h, v, h) \ . \tag{4}$$

   In this case, sequent 2 is an axiom. From sequent 1 the reasoning procedure generates the formula $\text{Cont}(z, I_z)$ and the computation process terminates. In conclusion, the generated property is: the function z is continuous in the interval I_z.
2. **Ob** is the conjunction of only two facts, i.e.:

$$\text{Cont}(u, I_u) \wedge \text{Cont}(v, I_v) \ .$$

   This time sequent 2 is not reduced to an axiom and the reasoning procedure generates $\text{Cont}(z, I_z)$ and abduces $\text{Int}(I_u, I_v, I_z) \wedge \text{Sum}(u, v, z)$. In this case the following conclusion is generated: "if the function z is the sum of the functions u and v and the interval I_z is the intersection of the intervals I_u and I_v then the function z is continuous in I_z."

The example described so far is relative to the functionality of the expert module, i.e., autonomous problem solving capability in the domain to be taught. As far as the other functionalities present in ITS (monitoring and diagnosis) are concerned, the proposed reasoning apparatus is capable to support them in a wide range of situations. For instance, suppose that the student makes a mistake in computing the sum of the functions. The system is able to recognize the mistake when it tries to match the student's answer with the result obtained by the application of the $\underline{\exists \mathbf{A}}$ rule (see Step 2). In another situation, suppose that the student reaches a wrong conclusion during her problem solving activity, e.g., given the observations (4), she performs correctly the required computations but she concludes that "the function z is not continuous in I_z". The system is able to detect such a mistake either by matching the student's solution with the (correct) solution built in advance by the expert module (if a specific problem has been proposed) or by verification of the validity of the student's assertion. The latter is the case when an "explorative" problem has been proposed, i.e., the system has given complete initiative to the student for exploring possible deductions of facts derivable in the given theory. In this case, the system builds the following sequent, obtained from (3), adding to its succedent the student's answer:

$$\begin{gathered}(\forall f, g, I_f, I_g \;\; \exists h, I_h) \\ ((\text{Cont}(f, I_f) \wedge \text{Cont}(g, I_g) \wedge \text{Int}(I_f, I_g, I_h) \wedge \text{Sum}(f, g, h)) \\ \supset \text{Cont}(h, I_h)), \mathbf{Ob} \rightarrow \neg\text{Cont}(z, I_z)\end{gathered}$$

The system tries to prove the validity of the above sequent using the proving part of the reasoning procedure. The result of the procedure is that the considered sequent is not satisfied. Thus the student's answer is incorrect. The diagnostic module requires an abduction (performed by the reasoning procedure) trying to determine the hypotheses under which the student's answer is correct. For brevity's sake, we do not go into many details of that procedure. The system builds the list of all the possible hypotheses. One of them is $\neg\text{Cont}(u, I_u)$, i.e., the function u is not continuous in its domain. In fact, if such a hypothesis were given, the student's answer should be true. Actually, if $\neg\text{Cont}(u, I_u)$ is a contradiction with respect to the given observations, then it has to be rejected.

5 Conclusions

In this paper a general reasoning apparatus for supporting various functionalities required in ITS for mathematics has been proposed. It is based on a sequent calculus, a proof system capable of actually solving mathematical problems, classified as verificative, generative, and abductive. It must be stressed the pedagogical prominence of such classes of problems, that are particularly stimulating for the students' intellect.

Sequent calculus is also able to be used in the natural deductive style, which closely mimics student reasoning. This feature allows a tutoring system to monitor the student during the teaching session, in the style of model-tracing ITS. In fact, during a teaching session, the system is able to monitor student actions step by step, check the exactness of the student answers and, in case of wrong or incomplete answer, pinpoint the mistake and give hints to bring him back on track. It can allow the student to take the initiative by asking for the correct next step, for a complete solution procedure, for suggestions or for explanations. We believe that with the proposed approach to the construction of ITS for mathematics, which includes both computing and reasoning capabilities in symbolic computation systems, we can realize new effective systems in terms of both problem solving capabilities and diagnostic power on the student's answers.

References

Aiello, L., Micarelli, A. (1990): SEDAF: an intelligent educational system for mathematics. Appl. Art. Intell. 4: 15–36.

Aiello, L., Colagrossi, A., Micarelli, A., Miola, A. (1993): Building the expert module for ITS in mathematics: a general reasoning apparatus. In: Nwana, H. S. (ed.): Mathematical intelligent learning environments. Intellect Books, Oxford, pp. 35–51.

Anderson, J. R., Boyle, C. F., Yost, G. (1985): The geometry tutor. In: Joshi, A. (ed.): Proceedings of the 9th International Joint Conference on Artificial Intelligence. Morgan Kaufmann, Los Altos, CA, pp. 1–7.

Bertoli, P., Cioni, G., Colagrossi, A., Terlizzi, P. (1997): A sequent calculus machine for symbolic computation systems. In: Miola, A., Temperini, M. (eds.): Advances in the design of symbolic computation systems. Springer, Wien New York, pp. 217–229 (this volume).

Bonamico, S., Cioni, G., Colagrossi, A. (1993): An enhanced sequent calculus for reasoning in a given domain. In: Miola, A. (ed.): Design and implementation of symbolic computation systems. Springer, Berlin Heidelberg New York Tokyo, pp. 369–373 (Lecture notes in computer science, vol. 722).

Brown, J. S., Burton, R. R. (1978): Diagnostic models for procedural bugs in basic mathematical skills. Cogn. Sci. 2: 155–192.

Burton, R. R., Brown, J. S. (1979): An investigation of computer coaching for informal learning activities. Int. J. Man Machine Stud. 11: 5–24.

Char, B. W., Geddes, K. O., Gonnet, G. H., Monagan, M. B., Watt, S. M. (1991): Maple V language reference manual. Springer, New York Berlin Heidelberg.

Cioni, G., Colagrossi, A., Miola, A. (1992): A desktop sequent calculus machine. In: Calmet, J., Campbell, J. A. (eds.): Artificial intelligence and symbolic mathematical computing. Springer, Berlin Heidelberg New York Tokyo, pp. 224–236 (Lecture notes in computer science, vol. 737).

Cioni, G., Colagrossi, A., Miola, A. (1995): A sequent calculus for symbolic computation systems. J. Symb. Comput. 19: 175–199.

Cioni, G., Colagrossi, A., Miola, A. (1997): Deduction and abduction using a sequent calculus. In: Miola, A., Temperini, M. (eds.): Advances in the design of symbolic computation systems. Springer, Wien New York, pp. 198–216 (this volume).

Limongelli, C., Miola, A., Temperini, M. (1991): Design and implementation of symbolic computation systems. In: Gaffney, P. W., Houstis, E. N. (eds.): Proceedings IFIP TC2/WG2.5 Working Conference on Programming Environments for High Level Scientific Problem Solving, Karlsruhe, Germany, Sept. 23–27, 1991. North-Holland, Amsterdam, pp. 217–226.

Moses, J. (1975): A MACSYMA primer. MathLab Memo no. 2, Computer Science Laboratory, Massachusetts Institute of Technology, Cambridge, MA.

O'Shea, T. (1982): A self-improving quadratic tutor. In: Sleeman, D. H., Brown, J. S. (eds.): Intelligent tutoring systems. Academic Press, London, pp. 309–336.

Sleeman, D. H. (1983): Inferring student models for intelligent computer-aided instruction. In: Michalski, R. S., Carbonell, J. C., Mitchell, T. M. (eds.): Machine learning: an artificial intelligence approach. Springer, Berlin Heidelberg New York Tokyo, pp. 483–510.

Sleeman, D. H. (1987): PIXIE: a shell for developing intelligent tutoring systems. In: Yazdani, M. (eds): Artificial intelligence and education: learning environments and tutoring systems. Ablex, Norwood, pp. 239–265.

Sleeman, D. H., Brown, J. S. (eds.) (1982): Intelligent tutoring systems. Academic Press, London.

Vivet, M. (1987): Systemes experts pour enseigner: meta-connaissances et explications. In: Proceedings Congres International MARI/COGNITIVA 87, Paris, France, May, 1987, pp. 18–22.

Wenger, E. (1987): Artificial intelligence and tutoring systems. Morgan Kaufmann, San Mateo.

Wolfram, S. (1991): Mathematica: a system for doing mathematics by computer, 2nd edn. Addison-Wesley, Reading, MA.

Yazdani, M. (1986): Intelligent tutoring systems: an overview. Expert Syst. 3: 154–162.

Subject index

Texts and Monographs in Symbolic Computation

Franz Winkler

Polynomial Algorithms in Computer Algebra

1996. 13 figures. VIII, 270 pages.
Soft cover DM 89,–, öS 625,–. ISBN 3-211-82759-5

The book gives a thorough introduction to the mathematical underpinnings of computer algebra. The subjects treated range from arithmetic of integers and polynomials to fast factorization methods, Gröbner bases, and algorithms in algebraic geometry. The algebraic background for all the algorithms presented in the book is fully described, and most of the algorithms are investigated with respect to their computational complexity. Each chapter closes with a brief survey of the related literature.

Jochen Pfalzgraf, Dongming Wang (eds.)

Automated Practical Reasoning

Algebraic Approaches

With a Foreword by Jim Cunningham

1995. 23 figures. XI, 223 pages.
Soft cover DM 108,–, öS 755,–. ISBN 3-211-82600-9

This book presents a collection of articles on the general framework of mechanizing deduction in the logics of practical reasoning. Topics treated are novel approaches in the field of constructive algebraic methods (theory and algorithms) to handle geometric reasoning problems, especially in robotics and automated geometry theorem proving; constructive algebraic geometry of curves and surfaces showing some new interesting aspects; implementational issues concerning the use of computer algebra systems to deal with such algebraic methods.
Besides work on nonmonotonic logic and a proposed approach for a unified treatment of critical pair completion procedures, a new semantical modeling approach based on the concept of fibered structures is discussed; an application to cooperating robots is demonstrated.

P.O.Box 89, A-1201 Wien • New York, NY 10010, 175 Fifth Avenue
Heidelberger Platz 3, D-14197 Berlin • Tokyo 113, 3-13, Hongo 3-chome, Bunkyo-ku

Springer-Verlag and the Environment

9 783211 828441